Assessment and Reclamation of Con

e retu
st d

ISSUES IN ENVIRONMENTAL SCIENCE AND TECHNOLOGY

EDITORS:

R.E. Hester, University of York, UK
R.M. Harrison, University of Birmingham, UK

TITLES IN THE SERIES:

FORTHCOMING:

How to obtain future titles on publication

A subscription is available for this series. This will bring delivery of each new volume immediately upon publication and also provide you with online access to each title via the Internet. For further information visit www.rsc.org/issues or write to:

Sales and Customer Care Department, Royal Society of Chemistry, Thomas Graham House, Science Park, Milton Road, Cambridge CB4 0WF, UK

Telephone: +44 (0) 1223 420066
Fax: +44 (0) 1223 423429
Email: sales@rsc.org

ISSUES IN ENVIRONMENTAL SCIENCE
AND TECHNOLOGY

EDITORS: R. E. HESTER AND R. M. HARRISON

16
Assessment and Reclamation of Contaminated Land

RS•C

ROYAL SOCIETY OF CHEMISTRY

ISBN 0-85404-275-X
ISSN 1350-7583

A catalogue record for this book is available from the British Library

Published by The Royal Society of Chemistry, Thomas Graham House,
Science Park, Milton Road, Cambridge CB4 0WF, UK

Registered Charity Number 207890

For further information see our web site at www.rsc.org

Typeset in Great Britain by Vision Typesetting, Manchester
Printed and bound by Bookcraft Ltd, UK

Preface

Despite a relatively stable total population, demographic trends in the United Kingdom are leading to an ever-increasing demand for housing. This, together with the requirements of industrial and commercial expansion, are leading to apparently insatiable demands for development land. If these demands are to be satisfied without yielding large tracts of countryside, development will need to make maximum use of brownfield sites, many of which will require remediation to render them suitable for their future purpose. *Issues* visited the subject of *Contaminated Land and its Reclamation* now almost five years ago in Volume 7, which has proved to be the most popular of all published volumes. Since that time a new regulatory framework for contaminated land has been introduced within the UK, the Environment Agency has developed a very active interest in the area and has stimulated new thinking on contaminated land risk appraisal, and there have been advances both in understanding the risks to health and the ecology posed by contaminated sites, and in remediation methodologies. We therefore felt it extremely timely to revisit the topic focussing especially upon these new developments.

The first chapter by Simon Pollard, Malcolm Lythgo and Raquel Duarte-Davidson of the Environment Agency's National Centre for Risk Analysis and Options Appraisal sets the scene by describing the basic problems of contaminated land as they affect the UK. The chapter summarises the origins and extent of problem and reviews some of the latest scientific developments that are underpinning effective contaminated land management. The following chapter by Malcolm Lowe of the Land Quality Team of the Department of Environment, Transport and the Regions and Judith Lowe, an independent consultant, on the new UK contaminated land regime sets out in some detail both the formal provisions of the new regulatory framework which was published for England during 2000 and its underlying rationale and intentions. One of the authors is a senior official in the government department responsible for this legislation, and the chapter provides a valuable insight into the philosophy of the new regime and the way in which it is expected to operate.

The volume then turns to more practical issues, and the next chapter by Mark Kibblewhite of the Environment Agency addresses the topic of risk assessment and risk management. Options for site investigation are described and problems with both sampling and chemical testing for contaminants are discussed. The importance of estimating the uncertainties of risk estimates is stressed as is the need for effective communication and project management. The main driver for much of the remediation of contaminated land is the concern over possible adverse effects on human health, and the chapter by Andrew Kibble and Pat Saunders of the University of Birmingham deals specifically with issues relating to human exposure to contaminated land and current evidence linking contaminated land with effects on public health. Sections include a discussion of the risk to children from soil ingestion, the use of biomarkers to demonstrate exposure to soil contaminants and evidence from epidemiological studies of the impact on public health. Having set the broad context of effects on health, the following chapter by Paul Nathanail and Naomi Earl of the Land Quality Management Group at the University of Nottingham describes the con-taminant–pathway–receptor approach to risk assessment and explains the scientific basis for quantification of risk and the derivation and application of guideline values for use in contaminated land remediation.

Not only human receptors are at risk from contaminated land. UK law offers protection also to environmental resources such as groundwater and ecosystems, and the sixth chapter, by Michael Quint of Parsons Brinckerhoff, addresses the issue of assessment of risks to ecological receptors. Key to this is assessment of the possibility of significant harm occurring. Moving to the more practical issue of remediation, Peter Wood of the University of Reading and AEA Technology, in a chapter on remediation methods for contaminated sites, reviews possible remediation strategies, highlighting advantages and disadvantages of different approaches and providing a summary of selected generic techniques. He then details a selection of innovative approaches that are starting to be applied within the UK. Emphasis is on remediation technologies that are essentially *in situ* techniques and which generally involve minimal disturbance of the ground surface. Finally, Anthony Lennon of ECS Underwriting describes the legal and financial minefields surrounding the issues of contaminated land and offers guidance on the restrictions regarding pollution liabilities in much insurance cover, highlighting specific underwriting approaches which can offer indemnity in such situations.

Overall, we believe that this collection of articles provides a comprehensive and detailed overview of the current state of the science of contaminated land. It has been contributed by authors authoritative in their own specialist fields and we commend it to those involved in the areas of land use planning and development, consultants and engineers in contaminated site investigation and remediation, environmental regulators, and to both teachers and students in higher education.

Roy M. Harrison
Ronald E. Hester

Contents

Issues in Environmental Science and Technology No. 16
Assessment and Reclamation of Contaminated Land
© The Royal Society of Chemistry, 2001

Contents

Editors

Ronald E. Hester, BSc, DSc(London), PhD(Cornell), FRSC, CChem

Ronald E. Hester is Professor of Chemistry in the University of York. He was for short periods a research fellow in Cambridge and an assistant professor at Cornell before being appointed to a lectureship in chemistry in York in 1965. He has been a full professor in York since 1983. His more than 300 publications are mainly in the area of vibrational spectroscopy, latterly focusing on time-resolved studies of photoreaction intermediates and on biomolecular systems in solution. He is active in environmental chemistry and is a founder member and former chairman of the Environment Group of the Royal Society of Chemistry and editor of 'Industry and the Environment in Perspective' (RSC, 1983) and 'Understanding Our Environment' (RSC, 1986). As a member of the Council of the UK Science and Engineering Research Council and several of its sub-committees, panels and boards, he has been heavily involved in national science policy and administration. He was, from 1991 to 93, a member of the UK Department of the Environment Advisory Committee on Hazardous Substances and from 1995 to 2000 was a member of the Publications and Information Board of the Royal Society of Chemistry.

Roy M. Harrison, BSc, PhD, DSc (Birmingham), FRSC, CChem, FRMetS, Hon MFPHM, Hon FFOM

Roy M. Harrison is Queen Elizabeth II Birmingham Centenary Professor of Environmental Health in the University of Birmingham. He was previously Lecturer in Environmental Sciences at the University of Lancaster and Reader and Director of the Institute of Aerosol Science at the University of Essex. His more than 300 publications are mainly in the field of environmental chemistry, although his current work includes studies of human health impacts of atmospheric pollutants as well as research into the chemistry of pollution phenomena. He is a past Chairman of the Environment Group of the Royal Society of Chemistry for whom he has edited 'Pollution: Causes, Effects and Control' (RSC, 1983; Fourth Edition, 2001) and 'Understanding our Environment: An Introduction to Environmental Chemistry and Pollution' (RSC, Third Edition, 1999). He has a close interest in scientific and policy aspects of air pollution, having been Chairman of the Department of Environment Quality of Urban Air Review Group and the DETR Atmospheric Particles Expert Group as well as currently being a member of the DEFRA Expert Panel on Air Quality Standards, the DEFRA Advisory Committee on Hazardous Substances and the Department of Health Committee on the Medical Effects of Air Pollutants.

Contributors

R. Duarte-Davidson, *Environment Agency, National Centre for Risk Analysis and Options Appraisal, Steel House, 11 Tothill Street, London SW1H 9NF, UK*

N. Earl, *Land Quality Management, School of Chemical Environmental and Mining Engineering, University of Nottingham, Nottingham NG7 2RD, UK*

A. J. Kibble, *Division of Environmental Health and Risk Assessment, University of Birmingham, Edgbaston, Birmingham B15 2TT, UK*

M. Kibblewhite, *Environment Agency, Land Quality Function, Rio House, Aztec West, Waterside Drive, Almondsbury, Bristol BS32 4UD, UK*

A. J. Lennon, *ECS Underwriting, IBEX House, 42–47 Minories, London EC3N 1DY, UK*

J. Lowe, *Independent Consultant, 160 Trafalgar Street, London SE17 2TP, UK*

M. Lowe, *Land Quality Team, Department of the Environment, Transport and the Regions, Ashdown House, 123 Victoria Street, London SW1E 6DE, UK*

M. Lythgo, *Environment Agency, Land Quality Function, Rio House, Aztec West, Waterside Drive, Almondsbury, Bristol BS32 4UD, UK*

C. P. Nathanail, *Land Quality Management, School of Chemical Environmental and Mining Engineering, University of Nottingham, Nottingham NG7 2RD, UK*

S. J. T. Pollard, *Environment Agency, National Centre for Risk Analysis and Options Appraisal, Steel House, 11 Tothill Street, London SW1H 9NF, UK*

M. Quint, *PB Environment, 4 Roger Street, London WC1N 2JX, UK*

P. J. Saunders, *Division of Primary Care, Public and Occupational Health, University of Birmingham, Edgbaston, Birmingham B15 2TT, UK*

P. Wood, *University of Reading and AEA Technology Environment, Postgraduate Research Institute for Sedimentology, PO Box 227, University of Reading, Whiteknights, Reading RG6 6AB, UK*

The Extent of Contaminated Land Problems and the Scientific Response

SIMON J. T. POLLARD, MALCOLM LYTHGO AND RAQUEL DUARTE-DAVIDSON

1 The Origins of Contaminated Land Problems

Background

The remediation and redevelopment of contaminated land is widely acknowledged as a major challenge for modern society.[1,2] Land, and the groundwater below it, used for siting industrial facilities or for waste disposal may be contaminated and its redevelopment often requires that unacceptable risks are assessed and managed so that the site becomes suitable for its new use. Over the years, development planning, corporate environmental auditing and environmental regulation have all contributed to the discovery and subsequent remediation of contaminated land and groundwater. Successful management of contamination, whether at the site, regional or national scale, relies on understanding and applying a large and multidisciplinary knowledge base that straddles the natural, physical, engineering and social sciences within a practical, commercial and regulatory context. Practitioners applying this knowledge work within a decision-making context far removed from that of the research environment. This said, practitioners and decision-makers are required to consider the significance of the science base that affects their decisions, particularly those aspects relating to the perception of risk, the environmental fate of contaminants, the capabilities of analytical science, the concepts of environmental exposure and the practical capabilities of remedial technologies.[3]

Decisions about contaminated land involve a multitude of stakeholders. New mechanisms and skills are required for engaging and involving audiences not familiar with the technically orientated language of environmental

[1] R. E. Hester and R. M. Harrison, eds., *Issues in Environmental Science and Technology*, No. 7, Royal Society of Chemistry, Cambridge, 1997.

[2] USEPA, United States Environmental Protection Agency, *Evaluation of Demonstrated and Emerging Technologies for the Treatment and Clean Up of Contaminated Land and Groundwater*, EPA542-C-99-002 (CD-ROM), Washington DC, 1999.

[3] S. E. Hrudey and S. J. Pollard, *Environ. Rev.*, 1993, **1**, 55.

Issues in Environmental Science and Technology No. 16
Assessment and Reclamation of Contaminated Land
© The Royal Society of Chemistry, 2001

risk.[4] The sustainable development policy agenda[5] is resulting in new ways of thinking about risk, technology and decision-making. Increasingly, approaches to site remediation are being scrutinised by reference to their full life-cycle costs, and social, economic and technical factors are being considered alongside one another in appraising risk management options.[6,7] Industry, its professional advisors and environmental regulators are now having to consider afresh how these broader aspects can best be incorporated into decisions.

A vast literature has developed on the issue of land contamination and several excellent reports, texts and reviews exist that focus on historical perspectives, policy and legal developments and consulting best practice.[1,8–13] Here, we introduce the origins and extent of the problem and, as a contribution to the debate over effective solutions to this environmental challenge, we highlight some of the recent developments in the science base that underpin management decisions on land contamination. Given the emphasis in later chapters, we have concentrated on the assessment of contaminated land and research in the open literature. Specific developments in risk assessment, remediation, impacts on human health and the policy and regulatory frameworks are provided in detail elsewhere within this volume.

Origins

The challenge of managing contaminated land is not a new one. It has been recognised by governments internationally for at least thirty years and is closely associated, technically and legislatively with the issues of waste and hazardous waste disposal, the regeneration of derelict land, groundwater pollution and

---

[4] Scotland and Northern Ireland Forum for Environmental Research, *Communicating Understanding of Contaminated Land Risks*, Project SR97(11)F, SNIFFER, Edinburgh, 1999.

[5] Department of Environment, Transport and the Regions, Cm4345, *A Better Quality of Life: A Strategy for Sustainable Development for the United Kingdom*, HMSO, London, 1999.

[6] R. P. Bardos, T. E. Kearney, C. P. Nathanail, A. Weenk and I. D. Martin, in *Contaminated Soil 2000*, Thomas Telford Ltd., London, 2000, pp. 99–106.

[7] S. J. T. Pollard, J. Fisher, C. Twigger-Ross and A. Brookes, in *Special Session of the NATO/CCMS Pilot Study of Demonstrated and Emerging Technologies for the Treatment and Clean Up of Contaminated Land and Groundwater*, Wiesbaden 2001, United States Environmental Protection Agency, Cincinnati, 2001, EPA 542-C-01-002 (CD Rom).

[8] House of Commons, Environment Committee, Session 1989–1990, *First Report: Contaminated Land*, HMSO, London, 1990.

[9] G. Fleming, *Recycling Derelict Land*, Institution of Civil Engineers, Thomas Telford Limited, London, 1991.

[10] P. Attewell, *Ground Pollution. Environment, Geology, Engineering and Law*, E. & F.N. Spon, London, 1993.

[11] J. Petts, T. Cairney and M. Smith, *Risk-Based Contaminated Land Investigation and Assessment*, John Wiley & Sons, Chichester, 1997.

[12] Construction Industry, Research and Information Association, *Remedial Treatment for Contaminated Land*, Special Publication SP101-112, CIRIA, London, 1998.

[13] C. Ferguson, D. Darmendrail, K. Freier, B. K. Jensen, J. Jensen, H. Kasamas, A. Urzelai and J. Vegter, eds., *Risk Assessment for Contaminated Sites in Europe, Volume 1, Scientific Basis*, LQM Press, Nottingham, 1998.

industrial site decommissioning.[14] Whilst legislative boundaries and specific definitions may apply to these activities separately, these issues continue to overlap to a large extent and an examination of the origins of the problem needs to reflect on the historical emergence and development of this set of issues on the socio-economic and political landscape.

The widespread reporting of individual incidents and near misses associated with land contamination, coupled with increased government interest in waste management during the 1970s and 1980s led in many countries to the development of new legislative controls dealing with historically contaminated land and preventing future land contamination (see chapter by Lowe and Lowe). Highly publicised and often referred to international incidents such as those at Love Canal in the United States and Lekkerkerk in the Netherlands[8] resulted in increased public and political interest. A growing concern over waste management practices together with a realisation of the practical and technical difficulties of redeveloping past industrial land have also fuelled the debate. In the United Kingdom, a watershed was the publication of the 1990 House of Commons Environment Committee report on contaminated land,[8] which itself had been prompted by previous Commons and House of Lords committee reports on toxic waste and hazardous waste disposal.[15,16]

The modern origins of the problem, however, begin with industrialisation, development of the fossil fuel-based economy and growth of the downstream chemical, heavy engineering and manufacturing industries from the 1800s onward (Table 1). Coal gas production and later petroleum refining, together with the steel and chemical industries (*e.g.* dyestuffs, wood preserving, chloralkali), whilst bringing benefits to society often left a legacy of environmental contamination in an era when a different philosophy prevailed over environmental protection. Table 1 lists some common contaminants and their related past industrial land uses and processes. On- and off-site disposal of industrial wastes, by-products and production residues was commonplace and in some circumstances considered, at the time, to be beneficial. For example, using process residues to raise land levels on floodplains created land for development with a reduced flood risk.

2 The Extent of the Challenge

The Nature of Contaminated Land Problems—the Issue of Definition

It is not a simple matter to define contaminated land, or to categorise degrees of contamination. Setting aside natural contamination, there is, in practice, a continuum of land contamination that results in a range of impacts, from the acute impacts of specific sites, through varying degrees of diffuse, more chronic effects of large areas, through to pristine, uncontaminated land. There is little

[14] P. J. Young, S. Pollard and P. Crowcroft, *Issues in Environmental Science and Technology*, No. 7, ed. R. E. Hester and R. M. Harrison, Royal Society of Chemistry, Cambridge, 1997, p. 1.

[15] House of Commons, Environment Committee, Session 1988–1989, *Second Report*: *Toxic Waste*, HMSO, London, 1989.

[16] House of Lords, Select Committee on Science and Technology, Session 1988–1989, *Fourth Report*: *Hazardous Waste Disposal*, HMSO, London, 1989.

Table 1 Potential contaminants related to selected industrial processes. A comprehensive list for a wide range of industries can be found in the DETR set of industry profiles[17]

Industry	Contaminant source/process	Example chemical
Gasworks/coking plants	Coal gasification	Polynuclear aromatic hydrocarbons (PAHs), Phenols
	Gas purification beds	Free cyanide/complex ferrocyanides, sulfides, sulfate
	Metals	Ni, Cu, Pb, Co, Cr, As
Oil refineries	Petroleum refining and downstream processing	Benzenes, PAH MTBE, V, sulfur
	Acids, alkalis	Sulfuric acid, sodium hydroxide
	Lagging, insulation	Asbestos
	Spent catalysts	Pb, Ni, Cr
Petrochemical processing/ production	As oil refineries	
	Metals	Cu/Cd/Hg
	Reactive monomers	Styrene, acrylate, vinyl chloride
	Initiators	Toluene di-isocyanate
	Amines	Aniline
Timber processing and treatment	Coal tar creosote	Phenol, xylenols, PAH
	Delivery agents (*e.g.* diesel)	Hydrocarbons, benzene
	Chlorinated phenols	Pentachlorophenol, mono-, di- and polysubstituted phenols
	Metals	As, Cu, Cr, Zn
Tanneries	Acid	Hydrochloric acid
	Metals	Cr^{III}
	Salts	Chlorides, sulfides
	Solvents	Kerosene, white spirit
Textile manufacture and processing	Metals	Zn, Al, Ti, Sn
	Acids, alkali	Sulfuric acid, sodium hydroxide
	Bleaching	Sodium hypochlorite
	Degreasing	Perchloroethylene
	Aromatic hydrocarbons	Phenols
	Pesticides	Dieldrin, aldrin, endrin
	Dyestuff residues	Cd, benzidine

land in the United Kingdom (UK) that has not been subject to some degree of contamination, albeit by long-range aerial deposition. There are differing views internationally on how much of this continuum should be considered as 'contaminated', by reference to pristine and/or background concentrations and the significance of the concentrations found. The UK has adopted an approach based on how suitable land is for its current or intended use. Land not suitable for

[17] Department of the Environment Transport and the Regions, *Industry Profiles*, DETR Publications, list at www.contaminatedland.co.uk/gov-guid/doe-pb.htm

Figure 1 The relationship between derelict land, contaminated land and land with contamination. Different terminologies have arisen due to the contexts (planning, regeneration, environmental protection) in which land is being considered for remediation. With enactment of Part IIA of the Environmental Protection Act 1990 the term 'contaminated land' has a statutory defintion[18]

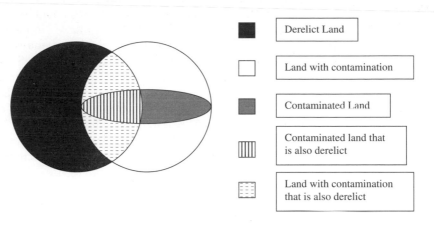

Derelict Land

Land with contamination

Contaminated Land

Contaminated land that is also derelict

Land with contamination that is also derelict

use is considered to be so contaminated that it needs to be remediated. This approach works well on a site-specific basis, but the definition can be problematic when trying to scale up for national estimates.

There are several terms in use to describe land in need of remediation. Although the terms may be similar, there are fundamental differences. Figure 1 illustrates the relationships between these terms. The term derelict land, for example, refers to 'land so damaged by past activities that it is incapable of beneficial use without treatment'.[14] The statutory definition of contaminated land in the UK is given in Part IIA of the Environmental Protection Act 1990 (see Lowe and Lowe), that is:

> "'Contaminated Land' is any land which appears to the local authority in whose area it is situated to be in such a condition, by reason of substances in, on or under the land, that
> (a) *significant harm is being caused or there is a significant possibility of such harm being caused*; or
> (b) *pollution of controlled waters is being, or is likely to be, caused*"

This statutory definition covers largely those sites at the chronically contaminated end of the continuum and relies on an assessment of the significance of the risks posed by the contamination at the site-specific level. An identified level of a particular contaminant may pose a significant risk of harm at one site with a sensitive land use, such as housing, but not at another less sensitive site, maybe where the land is used for parking. The same principle applies to the pollution of controlled waters, depending on the sensitivity of the receiving water. It might be argued that more general definitions would assist an understanding of the problem at the national scale, but generic descriptions without reference to the site-specific nature of the problem could undermine the risk-based approach and introduce blight on land values or saleability.

Extent and Significance

To date, there has not been a national survey of contaminated land, but there are

[18] Department of the Environment, Transport and the Regions, Circular 2/2000, *Environmental Protection Act 1990 Part IIA—Contaminated Land*, DETR, London, 2000.

provisions in recent legislation[18] for the environment agencies to report on the state of contaminated land in the UK and there is in hand work to produce a National Land Use Database (NLUD) that will provide estimates of the extent of derelict land. Current estimates of the extent of contaminated land in England and Wales range from 50 000 to 200 000 ha.[19] This is thought to represent approximately 100 000 contaminated sites, of which the Environment Agency estimates that between 5000 and 20 000 sites may require remediation because they pose an unacceptable risk to human health or the environment.[20] In 1998, the Welsh Office identified 949 contaminated sites in Wales, covering 4100 ha using a definition of contaminated land that excluded sites in active use, or those smaller than 0.5 ha.[21] In 1998, the Scottish Executive estimated that there were 3458 ha (52%) of derelict and vacant land that was known or suspected to be contaminated in Scotland and that the contamination status of a further 5553 ha was unknown.[22]

Estimates of the extent of contaminated land in the United Kingdom are often based on different definitions and therefore need to be viewed cautiously. They are often based on available but incomplete sources and they may mix between definitions. Further, some industry sectors have better information on the impact of their activities than others, and this might be misconstrued to imply that these sectors have caused more contamination. It is hoped that the provisions of the new contaminated land regime in the UK will refine and improve our estimates.

3 Developments in the Science Base

Developments in Environmental Science

From the first stirrings of political interest in contaminated land there has been a response from the scientific communities in terms of research output. Through the translation of research to the field, environmental science continues to underpin and contribute substantially to the practice of managing land contamination (Table 2). For example, our understanding of the fate and transport of contaminants in the environment has shaped our view on where and how to analyse for them, the exposure pathways associated with various classes of compounds and the technologies best suited to managing the risks posed.[3] The interdisciplinary nature of environmental science allows us to cross-fertilise between disciplines, for example, in applying analytical techniques used for petroleum exploration to the characterisation of oil contamination at sites or the application of process engineering design to remedial technology optimisation. Similarly, developments in the social sciences, not least with respect to risk perception and the communication of risk, are forcing a reappraisal of how we involve others in the process of managing land contamination.

A developing theme throughout the last 30 years has been the assessment, characterisation and management of risks to human health and the wider

[19] Royal Commission on Environmental Pollution, *Sustainable Use of Soils*, HMSO, London, 1996.

[20] Environment Agency, *The State of the Environment of England and Wales: The Land*, HMSO, London, 2000.

[21] Welsh Office, *Survey of Contaminated Land in Wales*, HMSO, London, 1998.

[22] Scottish Executive, *Scottish Vacant and Derelict Land Survey 1998*, ENV/1999/1, Scottish Executive, Edinburgh, 1999 available at: www.scotland.gov.uk/library2/doc05/lssb-00.htm

environment from contaminated sites.[13] Once it became clear that many environmental contaminants were ubiquitous at concentrations of parts per billion and less, a need developed to establish levels of significance for the potential risks posed by non-zero concentrations.[3] The risk assessment/risk management paradigm, adapted from the pharmaceutical and nuclear sciences, has since developed to a point whereby it has become a central point of reference for much of the supporting science and the basis for public policy and environmental regulation on contaminated land.[18] The application of risk assessment to contaminated land problems has inevitably highlighted issues of risk acceptance, the tolerability of risk and of equity.

Addressing Social Issues

Concerns over the potential risks to human health from exposure to contaminated soils have fired a broader debate over the wider social impacts of contaminated land on the quality of life. Many former industrialised areas form a focus for low cost or social housing, driven partly by positive issues such as access to work opportunities and partly by negative issues such as low land values. Some of the more deprived communities have resided in these contaminated areas. In addition to this social issue of environmental justice, a wider recognition of the importance of a sense of place and community attachment, and a broader interpretation of 'environment' beyond the strictly technical one, are bringing issues like the value of urban ecology, the importance of industrial heritage and the need for wider stakeholder involvement to the fore.

A move to more inclusive processes for environmental decision-making is emerging, influenced in part by 30 years of risk communication literature and by placing social progress centre stage in the strategy for sustainable development.[5] The regeneration of contaminated land has always required a multi- and trans-disciplinary approach, but increasingly scientists, engineers, planners and lawyers are turning to the social sciences for a re-interpretation of the issues historically viewed as driven by technological and economic concern alone. As a result, we are gaining valuable insights into the value of institutional trust, into 'process' issues in terms of involving others in decision-making, into issues of equity and the perceptions and reporting of risk. This work now needs to be integrated within the existing management frameworks for contaminated land management.[7]

Developments in Site Investigation and Assessment

The confidence with which decisions can be taken on the remediation of contaminated sites relies heavily on the quality of the site investigation and its attendant data. Risk assessments conducted in the absence of quality data provide a false sense of security in what can be meaningfully said about any site. Prioritising remediation across a 'portfolio' of sites requires comparability in data sets and summary information regarding the nature, extent, magnitude and depth of contamination. Developments in the statistical design of site investigations, in analytical science and the presentation of site data are all improving practice in this area.

Table 2 Selected developments and example references in environmental science of importance to contaminated land management

Example reference	Development	Significance of development
Alexander, 1965[23]	Emergence of soil microbiology and knowledge into biotransformations in soil/groundwater systems.	Development of bioremediation as a credible remediation technology. Optimisation of bioremediation technologies.
Schwille, 1967[24]	Developments in the characterisation of non-aqueous phase liquids (NAPLs) and their transport characteristics in the unsaturated and saturated zones.	Understanding of bulk flow and recovery of oil and solvents in porous media, sub-surface contaminant transport and contaminant dissolution from non-aqueous phase liquids.
Mackay, 1979[25]	Understanding of the fundamental physicochemical properties of contaminants as they dictate the environmental fate and transport of chemicals in the environment.	Better design of sampling protocols. Improved exposure assessment and understanding of remediation performance by reference to intra- and inter-phase transfer.
LeBlanc, 1984[26]	Development of commercially available computer software for the numerical simulation of contaminant transport in groundwaters.	Visualisation of plume fronts and behaviour. Predictive tools for natural attenuation and groundwater risk assessments.
Gauthier et al., 1987[27]	The characterisation of the soil matrix and, in particular, soil organic matter together with postulated models for intraparticle diffusion.	Understanding of fate and transport of hydrophobic contaminants in the soil environment and the thermodynamics and kinetics of release.
Paustenbach, 1987[28,29]	Codification of exposure models for estimating chronic intakes and characterising potential risks to human health from contaminated land.	Defensible basis for risk management. Targeting of risk management effort to principal exposure routes.

8

Swallow et al., 1988[30]	Commercially available analytical instrumentation, including coupled techniques (AAS; ICP; ICP-MS; HPLC; GC-MS) for the routine analysis of metals and organics in soils and groundwaters.	Reduced detection limits and rapid turnaround times for commercial laboratories undertaking contaminated land and groundwater analysis. Competitive market.
Chudyk, 1989[31]	Development and application of field analytical techniques for use in on-site characterisation.	Rapid, on-site analysis for directing detailed sampling regimes and on the spot decisions for soils shift operations.
USEPA, 1999[2]	Demonstration and evaluation of novel treatment technologies at the pilot and field scale. Practical examination of the interface between risk assessment and management.	Increased confidence in the transferability of laboratory-validated treatment techniques to the field. Identification of opportunities and constraints for individual techniques under heterogeneous conditions.
SNIFFER, 1999[4]	Application of risk communication research to issues of contaminated land.	Improved stakeholder dialogue and involvement in decisions regarding site investigation assessment and remediation.

23 M. Alexander, *Adv. Appl. Microbiol.*, 1965, **7**, 35.
24 F. Schwille, in *The Joint Problems of the Oil and Water Industries*, ed. P. Hepple, Elsevier, Amsterdam, 1967, pp. 23–54.
25 D. Mackay, *Environ. Sci. Technol.*, 1979, **13**, 1218.
26 D. R. LeBlanc in *Movement and Fate of Solute Transport in a Plume of Sewage-Contaminated Ground Water, Cape Cod, Massachusetts, United States Geological Survey Open File Report*, USGS, Virginia, 1984, pp. 11–45.
27 T. D. Gauthier, W. R. Seitz and C. L. Grant, *Environ. Sci. Technol.*, 1987, **21**, 243.
28 D. Paustenbach, *Comments Toxicol.*, 1987, **1**, 185.
29 D. J. Paustenbach, *J. Toxicol. Environ. Health (Part B)*, 3), 2000, 179.
30 K. C. Swallow, N. S. Shifrin and P. J. Doherty, *Environ. Sci. Technol.*, 1988, **22**, 136.
31 C. Chudyk, *Environ. Sci. Technol.*, 1989, **23**, 504.

Source Characterisation. Adoption of a risk-based approach to remediating land requires an understanding of the hazards posed by the source of contamination. Without these data, the development of practical solutions to remediation is fundamentally compromised. Soil contamination is invariably complex and heterogeneous in its distribution[32,33] and its volume is often difficult to estimate. This is a particular problem for non-aqueous phase liquids (NAPLs) that have migrated below ground level because they rarely flow vertically or laterally in a uniform manner.[24] However, recent developments in the combination of geotechnical and field analytical methods may offer site investigators increased confidence over their below-surface distribution.

Figure 2 illustrates results from a membrane interface probe cone penetration test at a trichloroethylene-contaminated site.[34] A standard cone penetrometer fitted with a hydrophobic membrane is advanced vertically below the ground using a truck-mounted hydraulic rig and the membrane heated as the probe advances. Volatile hydrocarbons passing the membrane flow to a gas chromatograph (GC) at the surface where they are quantified. Lithographic data are collected simultaneously and the data presented as a depth of contamination profile alongside the site geology. Selection of different detectors for the GC allows hydrocarbon and chlorinated solvents to be identified separately. Figure 2 includes data sets from electrical cone penetrometer (ECP) and membrane interface penetrometer (MIP) tests. Data from the photoionisation (PID), flame ionisation (FID) and electrolytic conductivity (DELCD) detectors are shown. The traces show the presence of mainly aromatic contamination in the top 4 m of the soil, with some chlorinated compounds in the top 11.5 m. The majority of the chlorinated compound is seen as dense (D)NAPL at 11.5–15 m below ground level. Without the DELCD, this distinction between the solvent types would not be realised. This novel approach has value both in informing the subsequent positioning of groundwater monitoring wells, for example, and in estimating the amounts of material requiring removal should this become necessary.

The interpretation of site investigation data, especially for large sites, requires statistical analysis. Many risk assessment tools require contaminant distributions or at least the input of worst-case and mean concentrations into deterministic exposure assessment models. The statistical design of site investigations and the representative sampling of contaminated soils has been reported on extensively. A central concern has been the collection of representative data sets that allow hot spots and total contaminated areas to be delineated and that provide randomised data that can be used in statistical risk assessment packages. Site assessments undertaken for the purposes of liability management have, by and large, adopted deterministic sampling approaches, concentrating effort at former process and waste disposal areas of the site by reference to the historical land use

[32] R.O. Gilbert, *Statistical Methods for Environmental Pollution Monitoring*, Van Nostrand Reinhold, New York, 1987.

[33] CCME, *Guidance Manual on Sampling, Analysis, and Data Management for Contaminated Sites, Volume 1: Main Report*, Canadian Council of Ministers of the Environment, Winnipeg, 1993, pp. 9–37.

[34] P.A. Jacobs, *Proceedings of the Conference on Implementation of In-Situ Remediation Techniques—Chlorinated Solvents and Heavy Metals*, Utrecht, October 9–10, 2000.

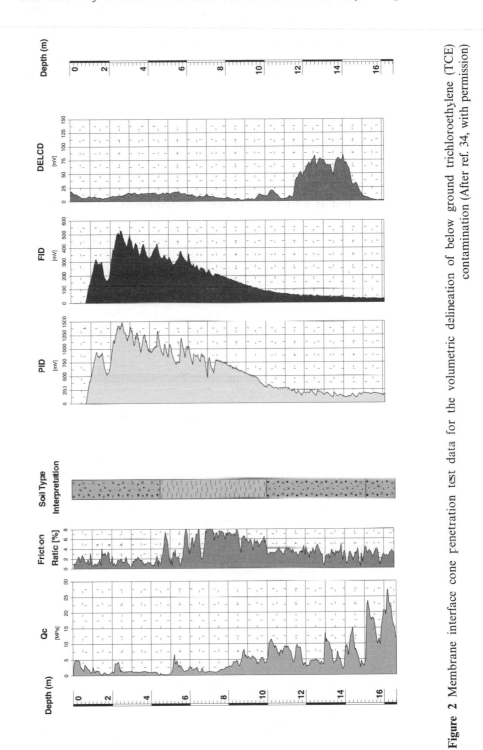

Figure 2 Membrane interface cone penetration test data for the volumetric delineation of below ground trichloroethylene (TCE) contamination (After ref. 34, with permission)

record. However, with the move to risk-based decision-making, the need to generate random, or at least stratified random, data sets on which statistical analyses can be performed is increasing. Where different needs are combined in the same site investigation, the separation of data sets to avoid the introduction of bias will be required.[33]

Analytical Methods. More powerful analytical instrumentation has become increasingly available and affordable over the last decades. Individual and coupled analytical techniques are now in widespread use and the routine availability of bench-top techniques has given greater access to more sensitive and selective measurements. Our appreciation of the complexity of contaminants in soil and groundwater systems is gradually driving a move to more sophisticated and informative analytical methods. The availability of improved techniques in itself has been an important driver towards a better understanding of these systems. Developments in risk assessment methods, which are increasingly concerned with specific and toxicologically relevant compounds[35] (see below), are driving changes in analytical method development, for example in the analysis of hydrocarbon residues in soils. The total petroleum hydrocarbon (TPH) risk assessment methodology[35] developed in the United States, for example, is requiring a new sophistication in the 'splicing' of the carbon number range in samples isolated from oil contaminated sites. Similar developments in our understanding of the toxicology of complex mixtures of contaminants (see below) have had implications with respect to the congener analysis of polychlorinated biphenyl- and dioxin-contaminated soils (Figure 3) and the analysis of polynuclear aromatic hydrocarbons (PAH) in soils.

In terms of metals analysis, the availability of inductively coupled plasma mass spectrometry (ICP-MS) is now providing better detection limits for important elements and wider simultaneous determinand suites than ICP-OES, though it is not reliable for the measurement of some contaminants such as mercury. Applied research into sequential extraction methods and physiologically based extraction techniques is informing our understanding of the role and importance of contaminant bioavailability in risk assessments.[29,36,37]

Detection limits continue to drop and legal disputes over cross boundary contamination are emphasising the need for quality assured data, inter-laboratory comparisons[38] and the development of standard methods for the purposes of analytical consistency (Figure 4). Alongside these developments in chemical analysis, biological diagnostic methods, such as enzyme-linked immunosorbent assays (ELISA) systems, are being increasingly used for rapid on-site analysis and

[35] D. J. Vorhees, W. H. Weisman and J. B. Gustafson, *Total Petroleum Hydrocarbon Criteria Working Group Volume 5: Human Health Evaluation of Petroleum Release Sites: Implementing the Working Group Approach*, Amherst Scientific Publishers, Amherst, 1999.

[36] S. E. Hrudey, W. Chen and C. Rousseaux, *Bioavailability in Environmental Risk Assessment*, CRC Press, Boca Raton, 1995.

[37] M. V. Ruby, R. Schoof, W. Brattin, M. Goldade, G. Post, M. Harnois and D. E. Mosby, *Environ. Sci. Technol.*, 1999, **33**, 3697.

[38] N. Boley, and D. Woods, *CONTEST Report, Round 27, March 2000*, Laboratory of the Government Chemist, Teddington, 2000.

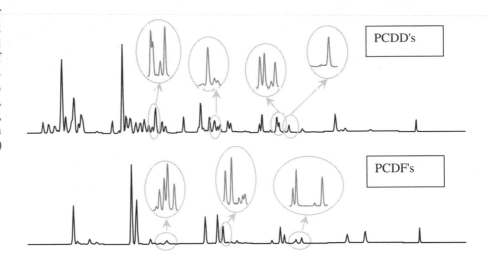

Figure 3 Analysis of 2,3,7,8-substituted congeners in soil; circled regions show sections of the traces separated by a second GC column. (Dr Nick Green, Environmental Sciences, Lancaster University, with permission)

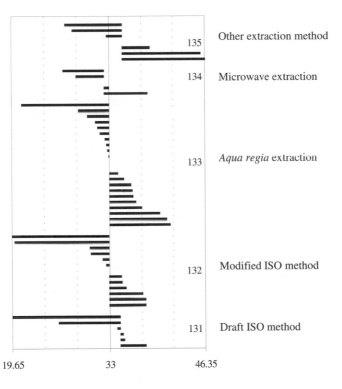

Figure 4 Distribution of analytical data for arsenic in a sample of homogenised soil, showing the performance of a range of laboratories (individual bars) and analytical methods (groups 131–135) about the grand median value of arsenic in the sample of 33 mg kg^{-1}. (After ref. 38, with permission)

to support the segregation and disposal of contaminated soils to on- and off-site facilities. The development and future application of lux-based (bioluminescent) bioassays for site characterisation, used alongside chemical analysis, offers site investigators the opportunity to evaluate and correlate the chemical mass and toxicity burden of polluted soils.

Figure 5a Three-dimensional section of modelled phenol plume in groundwater, showing the distribution of contaminant concentration (mol L^{-1}) away from the source area.
(After ref. 39, with permission)

Figure 5b Three-dimensional section of modelled phenol plume in groundwater, showing the distribution of contaminant degradation rates, with highest rates occurring at the plume fringe.
(After ref. 39, with permission)

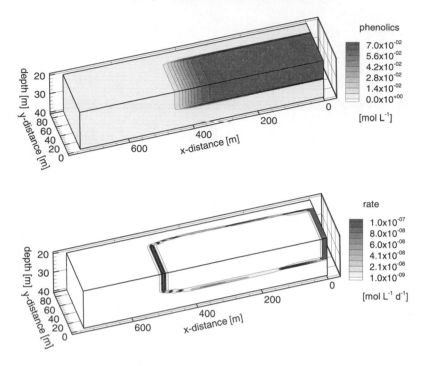

Visualisation Techniques. The development of numerical modelling packages and computer-based visualisation techniques for the simulation and presentation of spatial distributions of sub-surface contamination has greatly enhanced our understanding of the transport of bulk contamination. For example, these techniques allow the presentation and 'manipulation' of plume fronts[39] (Figures 5a and b) and inform estimates of travel times with or without the natural attenuation, including biotransformation, of substrates. Resources invested early in the interpretation of site data help refine detailed assessment and monitoring programmes and can lead to a greater understanding of site-specific effects, such as the relative importance of transverse diffusion or substrate biodegradation to plume advancement, for example. These techniques are now being routinely used to improve the quality of site decisions – from the design of supplementary site investigation to informing cost–benefit estimates of pump and treat strategies and post-remediation monitoring.

Risk Analysis

Many countries, including the UK, have adopted a risk-based approach to contaminated land management and use the familiar source–pathway–receptor approach to risk analysis (see chapter by Quint). Here, we highlight some of the recent contributions towards development of the risk assessment for contaminated land management.

[39] K. U. Mayer, S. G. Benner, E. O. Frind, S. F. Thornton and D. N. Lerner, *J. Contaminant Hydrol.*, 2001, *in press.*

Sources. No single analytical method is capable of measuring the range of contaminants present in the complex mixtures most commonly found at contaminated sites. Screening methodologies are now widely used for identifying key pollutants during an initial assessment of the site. Selecting the appropriate surrogates for screening is an important task and involves assigning toxicity and environmental behaviour criteria to well defined, inclusive groups of constituents based on toxicity studies completed on individual constituents, mixtures and whole products.

The TPH risk assessment methodology[35] approach above derives petroleum hydrocarbon 'fractions' with similar transport and fate characteristics, and assigns toxicity criteria to these fractions based on those of a representative 'surrogate' compound or, ideally, mixture. Using the actual mixture of concern allows for interactions such as synergism and antagonism that may occur among the constituents of the mixture, and changes in the relative mass of each fraction at the receptor to be accounted for in the risk assessment. Various working groups have contributed towards this debate by establishing appropriate surrogates for specific petroleum hydrocarbon fractions.[35,40–44] Examples of the fractions and toxicity effects derived by two of these working groups are summarised in Tables 3a and b. Screening approaches such as these offer a practical means of assessing the significance of potential health impacts from a chemically complex source. Based on the information obtained remediation or more detailed assessment of the site can be progressed.

Pathways. Exposure assessment models are continually being refined to take account of new scientific developments. There are now many commercially available tools for predicting the fate and transport of chemicals in the soil environment and for estimating the subsequent exposure of receptors to these contaminants through a variety of environmental pathways. Models only provide illustrative estimates of exposure, however, and require careful application on account of the numerous assumptions contained within. For example, the use of different contaminant transport models can lead to wide ranging predictions of exposure to specific contaminants with exposure point concentrations varying by orders of magnitude depending on the model selected.

Consider recent developments in our understanding of vapour intrusion into buildings. A number of soil vapour intrusion models have been developed including:

[40] American Society for Testing and Materials, *Standard Guide for Risk-based Corrective Action Applied at Petroleum Release Sites, Standard E1739-95*, ASTM, Philadelphia, 1996.

[41] M. S. Hutcheson, D. Pedersen, N. D. Anastas, J. Fitzgerald and D. Silverman, *Reg. Toxicol. Pharmacol.*, 1996, **24**, 85.

[42] Massachusetts Department of Environmental Protection, *Characterizing Risks Posed by Petroleum Contaminated Sites: Implementation of the MADEP VPH/EPH Approach. Draft for Public Comment*, MADEP, Boston, 1997, at: www.state.ma.us/dep/bwse/vph_eph.htm

[43] CONCAWE, *Environmental Risk Assessment of Petroleum Substances: The Hydrocarbon Block Method. Report No. 96/52*, CONCAWE, Brussels, 1996.

[44] Agency for Toxic Substances and Disease Registry, *Toxicological Profile for Total Petroleum Hydrocarbons*, ATSDR, Atlanta, 1999.

Table 3 Examples of hydrocarbon fractions, reference doses (RfDs) and reference concentrations (RfCs) recommended by two working groups[41,42,45-47] (a) TPHCWG and (b) MADEP

(a) *Hydrocarbon fractions, RfDs and RfCs from TPHCWG*

Carbon range (equivalent carbon number, EC)	Oral RfD (mg kg^{-1} day^{-1})	Inhalation RfC (mg m^{-3})
Aliphatic fractions		
EC_5–EC_8	5.0	18.4
> EC_8–EC_{16}	0.1	1.0
> EC_{16}–EC_{35}	2.0	—
Aromatic fractions		
EC_5–EC_7 (benzene)		
> EC_7–EC_8	0.2	0.4
> EC_8–EC_{16}	0.04	0.2
> EC_{16}–EC_{35}	0.03	—

(b) *Hydrocarbon fractions and RfD from MADEP*

	Reference compound	RfD (mg kg^{-1} day^{-1})
Aliphatic fractions (alkanes, cycloalkanes and isoalkanes)		
C_5–C_8	n-Hexane	0.06
C_9–C_{18}	n-Nonane	0.6
C_{19}–C_{32}	Eicosane	6.0
Aromatic fractions		
C_6–C_8	BTEXs	
C_9–C_{32}	Pyrene	0.03

- the Risk-Based Corrective Action (RBCA) Tool Kit;[48]
- the British Columbia (BC) model;[49]
- the British Petroleum RISC[50] (based on refs. 51 and 52);

[45] Total Petroleum Hydrocarbon Criteria Working Group, *Development of Fraction Specific Reference Doses (RfDs) and Reference Concentrations (RfCs) for Total Petroleum Hydrocarbons, Volume 4*, Amherst Scientific Publishers, Amherst, 1997, available at www.aehs.com/publications/catalog/contents/tph.htm

[46] Total Petroleum Hydrocarbon Criteria Working Group, *Selection of Representative TPH Fractions Based on Fate and Transport Consierations, Volume 3*, Amherst Scientific Publishers, Amherst, 1997, available at www.aehs.com/publications/catalog/contents/tph.htm

[47] Massachusetts Department of Environmental Protection, *Interim Final Petroleum Report: Development of Health-based Alternative to the Total Petroleum Hydrocarbon (TPH) Parameter*, MADEP, Boston, 1994, available at www.state.ma.us/dep/bwse/vph_eph.htm

[48] Groundwater Services Inc., *Guidance Manual for Risk-Based Corrective Action*, Groundwater Services Inc., Houston, 1995.

[49] I. Hers, R. Zapf-Gilje, R. Petrovic, M. Macfarlane and R. McLenehan, in *Environmental Toxicology and Risk Assessment (6th Volume)*, eds. F. J. Dwyer, T. Doane and M. L. Hinman, American Society for Testing and Materials, Philadelphia, 1997.

[50] BP Oil Europe, *BP RISC Software for Clean-Ups—Risk Assessment Software for Soil and Groundwater Applications, Version 3*, BP Oil Europe, Sunbury, 1997.

[51] P. C. Johnson and R. Ettinger, *Environ. Sci. Technol.*, 1991, **25**, 1445.

[52] P. C. Johnson, M. W. Kemblowski and R. L. Johnson, *Assessing the Significance of Subsurface Contaminant Vapour Migration to Enclosed Spaces—Site Specific Alternatives to Generic Estimates*, American Petroleum Institute, Publication Number 4674, Washington, 1998.

Table 4 Indoor concentrations of benzene measured and modelled (all values $\mu g\,m^{-3}$) after ref. 56

Measured concentration	GSI	BC diffusion	BC advection	RISC diffusion	RISC advection	Ferguson diffusion	Ferguson advection
1.6[a]	1.9	103	278	1.7	254	3.8	11
18[b]	760	7230	—	558	—	3.4E+04	—
44[c]	—	—	895	—	1270	—	7.6E+05
1479[d]	1470	—	2660	—	3660	—	2.8E+05

[a]Paulsboro data; [b]Chatterton data (diffusion); [c]Chatterton data (advection 30 Pa); [d]Chatterton data (steady-state advection 10 Pa).

- a model developed by Ferguson *et al.*;[53] and
- the VOLASOIL model.[54]

These models provide various representations of soil vapour intrusion following the migration into buildings of volatile organic compounds from a contaminated source. Differences in model predictions are caused by, amongst other aspects, differences in the processes being simulated and the assumptions made, the mathematical expressions used to simulate fate and transport and the practicality and reliability of obtaining and using input parameters required to run the models. As site-specific data become available, measured and predicted data can be compared and the information obtained used to refine these generic models.

Comparison of the processes simulated by these various models with sensitivity analysis performed on key input parameters (*e.g.* advection, soil moisture content, biotransformation rates) and calibration of the outputs from the models against site-specific data[55] can provide information on how each model performs against theoretical expectations and measured values (Table 4). This review showed that the predicted indoor air concentrations varied by over an order of magnitude (Table 1 in ref. 56). Incorporation of the key findings from this type of applied research work contributes towards improving the predictability of contaminants from contaminated land to various receptors.

Receptors. Developments have continued on the incorporation of human bioavailablity factors in human health risk assessment[37,57,58] and in the assessment of risks from contaminated land to ecological receptors.[59,60]

Dependent on the soil matrix and contaminant, only a fraction of the exposure

53 C. C. Ferguson, V. V. Krylov and P. T. McGrath, *Building Environ.*, 1995, **30**, 375.
54 M. F. W. Waitz, J. I. Freijer, P. Keule and F. A. Swartjes, *The VOLASOIL Risk Assessment Model Based on CSOIL for Soils Contaminated with Volatile Compounds*, National Institute of Public Health and the Environment (RIVM), Bilthoven, 1996.
55 Environment Agency, *Vapour Transfer in Soil Contaminants*, R&D Technical Report P309, Environment Agency, Bristol, 2001, *in press*.
56 D. Evans, I. Hers and R. Duarte-Davidson, in *Contaminated Soil 2000*, Thomas Telford Ltd., London, 2000, pp. 360–364.
57 R. R. Rodriguez, N. T. Basta, S. W. Casteel and L. W. Pace, *Environ. Sci. Technol.*, 1999, **33**, 642.
58 B. Smith, B. G. Rawlins, M. J. A. R. Cordeiro, M. G. Hutchins, J. V. Tiberindwa, L. Serunjog and A. M. Tomkins, *J. Geol. Soc.*, 2000, **157**, 885.
59 United States Environmental Protection Agency, *Guidelines for Ecological Risk Assessment*, EPA/630/R-95/002F, USEPA, Washington, 1998.
60 G. Suter, *Hum. Ecol. Risk Assess.*, 1999, **5**, 375.

point concentration of a contaminant may become bioavailable (for example, for ingestion exposures, absorbed through the gastro-intestinal wall, thereby entering the blood system and becoming available to the body's metabolic processes). Few risk assessments, with the exception of those for soil lead in the United States (US), attempt to address the human bioavailability of contaminants and they therefore tend to overestimate uptake. Recent research has focused on developing and validating *in vitro* laboratory test methods for estimating 'bioaccessibility' in soil[58] (the fraction of a substance that is released from the soil matrix when being digested in the human gastro-intestinal tract, thereby becoming available for absorption). Being able to represent and estimate bioaccessibility will be a major step towards assessing bioavailability. The findings of ongoing research will improve the scientific basis of estimating human health risks in soils and help support risk management decisions.

Ecological risk assessment for contaminated land has become a routine activity for site assessments in the US since the production of regulatory guidance in 1998[59] and is of growing interest in Europe on account of the EC Habitats Directive. The US has adopted a tiered, twin track approach that considers chemical exposure assessment alongside the analysis of ecological response. In practical terms, the measurement of ecological response has considered increased mortality rates at the population level, but the selection of appropriate end-points remains a contentious issue. Given the range of potential ecological end-points for characterising the significance of ecological harm, practitioners in the US have placed particular emphasis in the development and refinement of conceptual models[60] for ecological exposure and the characterisation of vulnerable or 'sentinel' species.

Effects.　Most risk assessments concentrate on individual chemicals, and there is a considerable challenge to develop approaches that consider the health effects caused by whole mixtures. Although still at early stages, approaches are being developed that account for the combined effects of mixtures. One example is the Toxic Equivalent Factor (TEF) approach, whereby the toxicity for each individual chemical in a group of chemicals related by their mechanism of toxic action is estimated based on its relative toxicity to the most potent chemical within the group. The toxic equivalent (TEQ) for a group of chemicals is estimated by (i) multiplying the concentration of each individual compound in the exposure medium (here, soil) by its TEF; and (ii) adding the individual TEQs to derive a total TEQ, assuming that the combined effect of all the compounds in the group is additive. This approach is commonly used to estimate the toxicity for 2,3,7,8-substituted dioxins, dioxin-like polychlorinated biphenyls (PCBs) and, to a lesser extent, those polynuclear aromatic hydrocarbons (PAHs) found to exhibit a similar toxicity mechanism to benzo[*a*]pyrene.

Direct toxicity assessment (DTA) can also be used to account for the combined effects of contaminants. Measurements taken from biological and ecotoxicological tests are more directly related to ecological risk, because they measure the biological effects from a mixture of contaminants and, in doing so, provide an implicit indication of the bioavailability of the contaminants. Some tests can be simple and cost-effective in comparison to screening for multiple chemical components. Lysosomal stability in earthworms, for example, has been used effectively to establish chemical exposure in contaminated soils, but more work is required to establish the cost-effectiveness of the technique and the relevance of results for predicting individual health.[61]

4   Conclusions

- The origins of contaminated land problems are well understood and relate largely to historical waste management practices in an era when a different environmental philosophy prevailed.
- Assessing the extent of the problem has been hampered historically by a lack of clarity over definitions. New legislation will help clarify the extent of land that is so contaminated that it poses a risk of significant harm to human health or the pollution of controlled waters.
- The scientific developments in this field have been considerable, especially with respect to environmental diagnostics, our understanding of the sub-surface, the advances made in the development of remediation technology and our understanding of the fate and transport of soil contaminants.
- A vast research literature has developed in the field of contaminated land management that continues to underpin policy and regulatory practice. The process of risk management has emerged as a central focus for much of this research. Several national and international research initiatives[62,63] have provided a basis for sharing expertise.
- Future research efforts are likely to continue along established paths and also seek to address the sustainable development agenda; focussing on the economics of contaminated land management, decision-making processes and sustainable technologies for remediation.

5   Acknowledgements

The authors wish to acknowledge the Environment Agency for permission to publish and Dr Mark Kibblewhite, Dr Jimi Irwin and Phil Crowcroft (Environment Agency) for valuable peer review comments received during its preparation. The views expressed are those of the authors alone.

[61] Environment Agency, *Review of Ecotoxicological and Biological Test Methods for the Assessment of Contaminated Land, R&D Technical Report P300*, Environment Agency, Bristol, 2000.

[62] S. J. T. Pollard and S. M. Herbert, in *Contaminated Soil '98*, Thomas Telford, London, 1998, pp. 33–42.

[63] Department of the Environment, Transport and the Regions, *Matching Contaminated Land Research to Need: Whose Research and Whose Need*, DETR, London, 1999, available at www.environment.detr.gov.uk/cll/index.htm

The New UK Contaminated Land Regime

MALCOLM LOWE AND JUDITH LOWE

1 Introduction

On 1 April 2000, the new regulatory system for contaminated land—Part IIA of the Environmental Protection Act 1990—came into force in England.[1] This was a significant milestone in the development of contaminated land policy in the United Kingdom, and the culmination of six years' explicit development of the regulatory regime itself. The regulatory system itself, however, forms only one part of a wider policy framework and regime for land contamination issues that has developed over a much longer period—perhaps as long as 25 years.

This chapter is intended to provide an introduction to the Part IIA system and an analytical view of its key dimensions, placing it in the context of wider policy on land contamination issues. (It does not provide a detailed, narrative description of the provisions of the Part IIA system—the system is described in detail in the circular published at the time the system came into force by the Department of the Environment, Transport and the Regions (DETR).[2])

This chapter covers:

- The context for Part IIA
- The Part IIA system
- 'Risk' and Part IIA
- Making people pay
- Part IIA as an 'regulatory system'
- The state of play
- What next?

2 The Context for Part IIA

The Context of the United Kingdom

Any country's contaminated land policy framework will have developed to reflect

---

[1] Separate arrangements are in place for Scotland, Wales and Northern Ireland. For simplicity, the references throughout this paper are specific to England.
[2] DETR Circular 2/2000 Environmental Protection Act 1990: Part IIA—Contaminated Land, March 2000.

Issues in Environmental Science and Technology No. 16
Assessment and Reclamation of Contaminated Land

both the intrinsic 'nature' of land contamination as an issue, and the 'nurture' of the policy as it developed in the context of particular priorities and other economic, legal and policy issues.

The United Kingdom has, of course, a long and varied industrial history. Most of the industrial sectors with the greatest potential to cause contamination problems have been active at some stage. Much of this industrial activity, particularly in the heavy industry sectors of coal extraction, iron and steel manufacture and chemicals production, took place long before modern systems of environmental controls on process emissions and waste disposal were in place. This, taken together with an equally long history of land recycling and the characteristic hydrogeological structures of the country, has shaped the type of contamination problems that we now face and the approaches adopted to deal with those problems.

A further key factor which has shaped the development of contaminated land policy has been the very active property market. This is particularly noticeable in the housing sector, where the UK has both a high proportion of owner-occupation and a rapid turnover of ownership in comparison with other countries. The ownership patterns of commercial land are also relevant. Much commercial land is held for investment purposes. But conversely in the case of smaller businesses, much of the financing of business operations is provided by short-term debt secured against the value of the property assets owned and occupied by the businesses. All of these factors have made questions of potential regulatory liabilities for land contamination politically and economically sensitive.

Overall Policy on Land Contamination

The Government's overall framework for policy on land contamination is set out in Annex 1 of DETR Circular 2/2000. Within the Government's overall objectives of achieving sustainable development, this policy approach has three essential components:

- Preventing the creation of new contamination;
- Promoting the remediation of the existing legacy of contamination through the redevelopment of land. This includes the provision of incentives such as financial assistance and grants; developing new technical solutions and promoting best practice; and the use of control systems, in particular building and development control; and
- Intervening through a regulatory process to deal with existing contamination where necessary.

For existing contamination, the Circular identifies three objectives:

"(a) to identify and remove unacceptable risks to human health and the environment;
(b) to seek to bring damaged land back into beneficial use; and
(c) to seek to ensure that the costs burdens faced by individuals, companies

and society as a whole are proportionate manageable and economically sustainable".[3]

There is an inter-play between these objectives. For example, dealing only with *'unacceptable* risks' restricts the costs that are involved. This helps to enable more land to be reclaimed for new uses within limited resources available either to private investors or to public regeneration agencies.

The objectives are also brought together in the 'suitable for use' approach. Under this approach, the potential risks presented by any site, and therefore the decision as to whether remediation is necessary, are assessed on the basis of the particular use of the land and its environmental setting. (The 'environmental setting' encompasses issues such the underlying geology and hydrogeology of the site, as well as potential impacts on neighbouring sites and important natural habitats.)

The particular role of the Part IIA system is to ensure that land is 'suitable for use' in the context of its current use and setting. Previous statutory regimes have been in place for similar purposes—most noticeably and recently, the 'anti-pollution works' provisions in the Water Resources Act 1991[4] and the 'statutory nuisance' system in the Environmental Protection Act 1990.[5] But following a policy review in 1993 and 1994,[6] the then Government considered that a 'modern, specific contaminated land power' was needed,[7] and legislated accordingly in the Environment Act 1995.

Circular 2/2000 describes the current Government's primary objectives for introducing the new regime as being:

'(a) to improve the focus and transparency of the [existing] controls, ensuring authorities take a strategic approach to problems of land contamination;
(b) to enable all problems resulting from contamination to be handled as part of the same process; previously separate regulatory action was needed to protect human health and to protect the water environment;
(c) to increase the consistency of approach taken by different authorities; and
(d) to provide a more tailored regulatory mechanism, including liability rules, better able to reflect the complexity and range of circumstances found on individual sites'.[8]

A further essential ingredient at both stages has been the desire to underpin voluntary remediation, whether in the context of redevelopment or pre-empting formal regulatory intervention. For both, the reduction in uncertainties surrounding so-called 'residual liabilities'—in terms of both extent and allocation—has been a major driver for the development of Part IIA.

---

[3] Paragraph 7 of Circular 2/2000, Annex 1.
[4] Section 161, Water Resources Act 1991.
[5] Part III, Environmental Protection Act 1990.
[6] *Paying for our Past*, Department of the Environment and Welsh Office, 1994.
[7] *Framework for Contaminated Land*, Department of the Environment and Welsh Office, 1994.
[8] Paragraph 26, Circular 2/2000, Annex 1.

3 The Part IIA System

Part IIA

The Part IIA system is set out in three key legal documents:

- *Part IIA of the Environmental Protection Act 1990.*[9] This 'primary legislation' was inserted into the 1990 Act by section 57 of the Environment Act 1995, and consists of new sections 78A to 78YC.
- *The Contaminated Land (England) Regulations 2000.*[10]
- *Statutory Guidance issued by the Secretary of State.* This is set out in Chapters A to E of Annex 3 of DETR Circular 2/2000.

There is a strict hierarchical relationship between these three elements, with the primary legislation at the top. The regulations and the guidance are issued under powers created by the primary legislation, and their scope is restricted to dealing with detailed issues specified by that legislation. Regulations and the guidance cannot contradict a provision of the primary legislation; nor can they create new provisions within the system as a whole for which the primary legislation does not provide specific legal powers, or *vires*.

There are four other legal documents which are also relevant to the implementation of Part IIA. These are all 'statutory instruments'—that is, Parliamentary orders or sets of regulations. First, there were two routine commencement orders bringing the provisions of the primary legislation into effect. These are described in Annex 5 of DETR Circular 2/2000. Perhaps more interesting are the other two:

- *Pollution Prevention and Control (England and Wales) Regulations.*[11] This does two things to Part IIA: first, it amends the provisions in the Part IIA primary legislation preventing overlap of regulatory functions, building in a cross reference to the new Pollution Prevention and Control regime; and secondly, it amends the Contaminated Land Regulations to add a new category of special site and to ensure that regulatory action taken under the PPC regime on contaminated land is recorded in the Part IIA registers.
- *Contaminated Land (England) (Amendment) Regulations 2001.*[12] This inserts a reference to the newly created concept of 'justices' chief executives' into the regulation governing the handling of appeals in the magistrates' court.

(Legislation does change over time, and it is always advisable to look at an up-to-date consolidated text in a legal encyclopaedia, rather than to rely solely on the text of legislation as originally published by the Stationery Office. Amendments made over time may not always be hugely significant in terms of the broad thrust of policy, but they could be critical on an individual site.)

---

[9] Environmental Protection Act 1990, as amended by Environment Act 1995.
[10] The Contaminated Land (England) Regulations 2000, SI 2000/227.
[11] Pollution Prevention and Control (England and Wales) Regulations 2000, SI 2000/1973.
[12] Contaminated Land (England) (Amendment) Regulations 2001, SI 20001/663.

The 'Content' of Part IIA

A definition of 'contaminated land':

'Contaminated land' is defined for the purposes of Part IIA by section 78A(2). This definition is then amplified by the statutory guidance in Chapter A of Circular 2/2000, which covers both the manner in which individual local authorities determine whether any land meets this definition and the interpretation of the concepts of 'significant harm' and the 'significant possibility' of significant harm. (These are explored in more detail later in this chapter.)

Potential problems arising from land contamination have previously been addressed by UK law. For example, Section 161 of the Water Resources Act 1991 refers to situations where "any poisonous, noxious or polluting matter or any solid waste matter is likely to enter . . . any controlled waters", which clearly includes the sense of one type of land contamination problem. The concept of contaminated land has also been referred to in UK law—for example, the Leasehold Reform, Housing and Urban Development Act 1994 gave the Urban Regeneration Agency (which became known as English Partnerships) specific powers to fund the reclamation of derelict, contaminated or un-used land. But Part IIA contains the first definition of contaminated land in UK law.

Other Approaches to Definition. There have been other working definitions in policy and practice, but specific legal definitions are of course only developed at the point at which a definition is needed to drive some particular statutory process. The definition will then be focussed on the particular needs of that process. The Part IIA contaminated land definition is quite deliberately restricted to the categories of site that Part IIA is intended to deal with—that is, sites that are currently presenting unacceptable risks to human health and the wider environment. The definition describes land where the contamination is not just present, but where it is, or could be, causing a particular problem.

The definition does not cover all land that contains some degree of contamination, which might be the 'man on the street's' definition of contaminated land; it also does not include all land that would fall within many of the technical descriptions of contaminated land, such as that proposed by the NATO Committee on Challenges to Modern Society (CCMS) pilot study on innovative remedial technologies. The general description 'land affected by contamination' has been used by DETR and now others to put a name to the wider categories of land which are dealt with in UK policy or legislation in different ways.

Other Approaches to Terminology. Some commentators have suggested that Part IIA should have adopted a term other than 'contaminated land'—such as 'polluted land'—to describe the particular category of 'contaminated land that requires regulatory intervention', and thereby to remove any confusion over what the term is meant to imply. This was specifically rejected in the drafting of Part IIA, not least because it raised various semantic drafting arguments in particular about the term 'polluted'—did it, for example, pre-suppose some positive act of polluting? But public debate over contaminated land policy had also heightened perception over 'contaminated land', and the term was increasingly being used to describe problem sites, rather than just sites with contamination present but harmless.

In addition, across Europe the terms 'polluted' and 'contaminated' are to a large extent synonymous, again with 'contaminated land' becoming more generally used as the term to describe problem sites. The intention in the UK was certainly not to imply that the scope of Part IIA was wider than it actually is, but rather to focus debate on when land contamination is a problem requiring regulatory intervention and what form that intervention should take.

Duties on regulatory authorities: An important feature of the overall design of the Part IIA system is that its key mechanisms for finding and dealing with problem sites is based on two linked 'duties' placed on regulatory authorities. (Legally, 'duties' are typically phrased as things that 'shall be' done—'powers', however, are described as things that 'may be' done, and their use is discretionary. Case law established under a very similar duty to take enforcement action, in the statutory nuisance system, has demonstrated that a duty is something the authority *has* to do; it cannot hold back just because it does not want to take action, for whatever reason.)

The first of these duties is to take action to identify contaminated land. This duty falls on a borough, district or unitary council, which is to 'cause its area to be inspected from time to time for the purpose of . . . identifying contaminated land'.[13] Local authorities therefore have a duty to try to find problem sites; this duty is given more emphasis in the accompanying statutory guidance which requires each authority to adopt a strategic approach to the process, describing the approach it will adopt in a published document. It will therefore be hard for any authority to expect, legally, to sit on its hands and deal with any explicit complaints about sites made by members of the public or local businesses. Some kind of positive action will be necessary.

The second of the duties is to 'require remediation' of any land identified as meeting the definition of contaminated land. This takes the form of a duty to serve a remediation notice[14] unless particular circumstances apply. In effect, these circumstances, which are set out in detail in the legislation, fall into four categories:

- No remediation is appropriate, in the light of a cost–benefit assessment;
- Enforcement action is being, or should be, taken under other regulatory powers;
- The authority is satisfied that the necessary remediation will take place without a remediation notice being served; or
- The Part IIA legislation does not permit the imposition of the full cost of the remediation as a liability on one or more persons.

This duty to require remediation falls on the 'enforcing authority' for the particular piece of contaminated land. In most cases, this will be the local authority that identified the land as being contaminated land. But Part IIA also creates a sub-set of contaminated land called 'special sites', for which the Environment Agency takes over the role of enforcing authority. Here again, there

---

[13] Section 78B(1), EPA 1990.
[14] Section 78E(1), EPA 1990.

is no room for discretion on what is, or is not, a special site. Regulations prescribe particular descriptions of contaminated land which are *'required to be designated a special site'*[15] (emphasis added).

Although the general pattern for the remediation process is for the enforcing authority to be under a duty to act, the legislation carefully moves to a power to take action—and is therefore discretionary—in circumstances where the burden of paying for the work falls on the enforcing authority. So in the fourth of the cases outlined above where a remediation notice cannot be served the enforcing authority acquires a power to carry out the remediation itself.

Mechanisms to enforce remediation requirements: The Part IIA primary legislation, statutory guidance and regulations set out a detailed machinery for defining and enforcing remediation requirements. This covers a range of practical issues related to enforcement, including remediation standards, the timing for the service of remediation notices, and provisions ensuring that those required to carry out remediation on land they do not control have the legal right to carry it out (with compensation being payable to those having to grant the necessary rights of access).

Rules on liabilities: Part IIA provides what is, in international terms, probably the most detailed framework of 'rules' for determining questions of liabilities for contaminated land remediation. International experience shows, at a simple level, that without a clear structure of rules on liabilities to pay for remediation, remediation generally does not happen—those who are not certain if they will become liable lose any incentive to pre-empt enforcement, any enforcement action becomes difficult or impossible, and public budgets are never sufficient to pay for the remediation.

At one level, the decision in the UK to provide a large amount of detail merely reflects the complexity of the issues to be resolved. These were spelled out in the Government's policy review during the early 1990s.[16] But it also reflects the fact that this policy review and the subsequent decision to introduce Part IIA were to a significant extent driven by concerns over liability issues, and the perceived need to define potential liabilities in close detail in order to remove regulatory blight from the property market. This dimension is discussed later in the chapter, as are many of the key features of the liability rules themselves. The main 'rules' themselves are set out in Section 78F (which is filled out in statutory guidance in Chapter D of Annex 3 of DETR Circular 2/2000), with subsidiary rules and provisions for special cases being set out in Sections 78J, 78K, 78P and 78X.

Provisions on the recording and release of information: In line with general practice for enforcement regimes, Part IIA includes requirements for authorities to record information about their enforcement actions on publicly-accessible registers.

Unless an enforcing authority effectively changes its mind about a decision that a particular piece of land is 'contaminated land', its formal identification will lead to information about that land being recorded on a public register. Remediation

---

[15] Section 78C, EPA 1990; the relevant 'descriptions' are set out in regulations 2 and 3 of the Contaminated Land (England) Regulations 2000.

[16] *Paying for our Past*, Department of the Environment and Welsh Office, March 1994. *Framework for Contaminated Land*, Department of the Environment and Welsh Office, November 1994.

notices—which require someone other than the enforcing authority to carry out specified remediation—are recorded. Where remediation is taking place 'voluntarily'—including where the enforcing authority is acting itself—a 'remediation statement' has to be prepared and placed on the register describing the intended work. In cases where no remediation is appropriate, a 'remediation declaration' records this and the reasoning behind the decision. These requirements potentially provide a strong tool for accountability of the enforcing authority—their decisions have to be made public.

The whole question of public information about land contamination is fraught with difficulties and concerns about potential 'blight' (or, to be more accurate, 'stigma'). The property market, in particular, has been strongly sensitised to these issues not least by its own reactions to the 'Section 143 registers' proposals in the early 1990s.[17] The enforcement registers under Part IIA are intended to be a different kind of mechanism, linked to regulatory good practice: effectively, they record details of the remediation work that is in progress, or has already happened, to make sites safe, rather than being a waiting list of sites about which there are still strong concerns or, more softly, grounds for suspicion.

The difference is not so much in the content of the registers—the consultation paper on the proposed Section 143 registers[18] also proposed that information on actual risk assessment and remediation that had been carried out would be included on the register as well as earlier site history—but on the point in the assessment process at which the sites get recorded. Section 143 would have cut in early, leading to a large number of 'false positive' identifications if the focus is only on current 'problem sites'. A Part IIA register entry, by contrast, is not made until a site has been *both* formally assessed and identified as 'contaminated land' *and* either a remediation notice, statement or declaration has been produced or the site has been designated a special site.

Removing Sites.   The treatment of cases once they have been remediated under Part IIA has been a focus of particular concern by some lobbyists. They argued that these sites should be removed from the register because they would no longer be 'contaminated land'. This suggestion has, however, been consistently and strongly resisted by DETR.

First, there would be no other reliable means of recording what has happened on a site—an authority would need to maintain some form of record for a reasonable timescale, and it is sensible to make this the managed public record of regulatory action set up under Part IIA. The Environmental Information Regulations would, in any event, mean that these kinds of records would be open to public scrutiny.

Perhaps more importantly though, "remediation" under Part IIA and the 'suitable for use' approach is in essence a process of managing and controlling risk in very specific circumstances. In particular, a site that has been made safe in the context of a particular land use may require further remediation if a new, and more sensitive, land use is proposed. It therefore seems entirely appropriate that

[17] Section 143 (now repealed), EPA 1990.
[18] Public registers of land which may be contaminated—a consultation paper; Department of the Environment and Welsh Office, May 1991.

once information about the land has been 'captured' by public authorities, this information should remain accessible so that everyone involved can be suitably appraised of the true situation.

'Signing-off' Remediation.   A similar logic affects the question of 'signing off' remediation. Part IIA does not contain any explicit 'signing off' provisions—it is not practicable to provide a process that might be taken as a publicly-backed guarantee that the site was now 'safe', especially as this might readily be misinterpreted as being for all time and in all circumstances. The Part IIA registers do allow those responsible for any remediation—or for the land itself—to place their own information about the remediation on the public record. But this is on the basis that the inclusion of this information on the register does not imply any endorsement of its accuracy by the authority holding the register.

Of course regulatory authorities will have to develop systems for checking that the notice has been complied with, and for taking action if it has not. An authority can then confirm whether or not it proposes further action on a site such as a prosecution for non-compliance with a remediation notice or the service of a further remediation notice. The absence of further action implicitly carries with it some reassurance that the remediation has been carried out effectively.

4   Dealing with 'Risk'

The Concept of 'Risk'

The notion of managing environmental 'risk' lies at the heart of Part IIA. The overall 'risk-based' nature of the system implies that it is anticipatory (it does not have to wait for the actual manifestation of harm) and that it is not arbitrary in its approach (for example, by focussing regulatory action solely on particular past land uses or at some particularly point in time, such as when the land is being sold). But the concept of risk, and risk management, permeates throughout Part IIA—in the detailed text, in the overall process, and in the choices that are made in the system.

Risk in the Definition of 'Contaminated Land'

Risk is defined in the statutory guidance as "the combination of:

(a) the probability, or frequency, of occurrence of a defined hazard (for example, exposure to a property of a substance with the potential to cause harm); and

(b) the magnitude (including the seriousness) of the consequences".[19]

The most immediately apparent area where this concept of risk drives the Part IIA system is in the core definition of contaminated land, set out in Section 78A

---

[19] Paragraph A.9, Circular 2/2000, Annex 3; although this definition was, in turn, borrowed from wider DETR publications on environmental risk—see later footnotes.

and the statutory guidance in Chapter A of Annex 3 of Circular 2/2000.

'Contaminated land' is defined for the purposes of Part IIA by Section 78A(2). This states that:

> "'Contaminated land' is any land which appears to the local authority in whose area it is situated to be in such a condition, by reason of substances in, on or under the land, that—
>
> (a) significant harm is being caused or there is a significant possibility of such harm being caused; or
> (b) pollution of controlled waters is being, or is likely to be, caused".

This clearly introduces the idea of 'possibility' or likelihood which can be related to the scientific terms 'probability' or 'frequency'. It also builds in the concept of 'consequences'—*i.e.* 'significant harm' or 'pollution of controlled waters'.

The 'Pollutant Linkage' Concept. A further concept within the definition—the 'contaminant–pathway–receptor' approach—is elaborated in the statutory guidance. This concept has, in recent years, come to represent the predominant intellectual framework for risk assessment in general[20,21] and has underpinned the development of contaminated land technical thinking. Unless a particular receptor could be harmed, through a defined pathway, by an identified contaminant, then land cannot be considered to meet the definition of contaminated land.

The statutory guidance introduces the term 'pollutant linkage' to describe an individual 'contaminant–pathway–receptor' relationship. This is key not only to the process of evaluating particular risks, but also to characterising a site as a whole—which may have many pollutant linkages—and thereby driving the remediation and liabilities aspects of the regulatory regime.

Statutory Receptors. The main components of the statutory guidance in Chapter A deal with the questions of what harm is to be regarded as 'significant', and whether 'the possibility of significant harm being caused is significant'. The answers to these questions are given in two tables: Table A sets out categories of 'significant harm'; and Table B describes what would be a 'significant possibility' with respect to each of these categories.

Table A, in particular, represents detailed policy choices on what types of risk ought to trigger regulatory intervention. It identifies four categories of harm based on specific receptors—which have been labelled in some of the secondary literature as 'statutory receptors', and the descriptions of the type or level of harm within each of these which would be significant.

The first category relates to 'human health effects'. This is relatively straightforward (leaving aside complications relating to the previously largely

---

[20] *A Guide to Risk Assessment and Risk Management for Environmental Protection*, Department of the Environment, 1995.
[21] *Guidelines for Environmental Risk Assessment and Management*, DETR, Environment Agency and Chartered Institution of Environmental Health, 2000.

uncharted territory on the issue of mental dysfunctions).

The second category of 'ecological system effects' is more complex. The basic definition of 'harm' in the primary legislation includes harm to the health of any living organism or other interference with the ecological systems of which they form part. If unqualified by the statutory guidance, this could have represented a 'zero tolerance' approach to the implications of any land contamination. This approach was considered by the Government, and more importantly by Parliament, to be impossible to achieve in practice and unnecessary.

The statutory guidance therefore narrows the scope to locations that are already protected under various legal habitat and site designations. It then further qualifies this further by setting a standard for the amount of harm that, in effect, would compromise the particular features of the location that led to its original designation. For example, for 'European Sites'—such as those in the Natura 2000 network—this is translated into 'harm which is incompatible with the favourable conservation status of natural habitats at the location or species typically found there'. This direct link between EU directives on nature conservation and contaminated land legislation is new within Europe, although it may become in effect an obligation under the proposed directive on environmental liability.

The third and fourth categories of receptor relate to property damage—'animal or crop effects', and 'building effects' respectively. Here again, the explicit identification of these forms of potential harm within the same statutory framework, particularly with respect to buildings, has tried to anticipate the development of international thinking.

The Approach to Water Pollution.   It is important to note that the definition in Part IIA treats any 'controlled waters' as a receptor in their own right—not just as a pathway or medium for other receptors. This reflects both national and European policy objectives to protect water as a vital environmental resource.

In this intellectual context, the treatment of water pollution in Part IIA can be seen to be risk based. But due to the particular way in which the primary legislation is drafted in this respect, it seems to be triggered by almost any detectable degree of risk; it is potentially much more stringent than the treatment of other receptors, in that there is no 'significance' test of the magnitude of the consequences. This is in line with definitions in other legislation on protection of controlled waters.

DETR has suggested, in Annex 2 of Circular 2/2000, that in practice what it calls "very slight levels of water pollution" are unlikely to lead "to the imposition of major liabilities". In particular, this is because any such water pollution would be unlikely, by definition, to have any degree of seriousness. Therefore an enforcing authority, required to consider what remediation would be 'reasonable' having regard to its cost and the seriousness of the problem it is intended to address, would not be able to justify any substantial cost.

However, DETR did commit itself in the Circular to seeking amendments to the legislation on this point to cover the specific test for contaminated land under Part IIA. The first opportunity to meet this commitment may be presented in a forthcoming Water Bill; a consultation paper on a draft bill was published in

November 2000. This included the proposal to amend the definition of contaminated land to refer to 'significant pollution of controlled waters' and the 'significant possibility of such pollution'. As with the existing treatment of 'harm', statutory guidance would be issued on what is significant.

A 'Risk Management' Process

Over and above the way that the definition of contaminated land implements a risk-based approach, the whole Part IIA process adopts a 'risk management' approach, with explicit considerations of risk being the key drivers at each stage.

The overall process within Part IIA broadly follows five key stages recognisable from an overall framework for dealing with environmental risks:

- Screening (strategic inspection);
- Detailed assessment of risks (determination of whether land meets the definition);
- Options appraisal (selection of remediation approach);
- Risk management (securing remediation and recording action); and
- Data collection and monitoring (strategy review, national report).

These stages are discussed below using the headings 'risk assessment' (the first two stages) and 'risk management' (the later stages); these headings mirror the two key duties on enforcing authorities to identify problem sites and to require their remediation.

Risk Assessment

The statutory guidance emphasises that the first duty of local authorities—to "cause its area to be inspected . . . to identify contaminated land"—is to be carried out in a risk-based way. The key stages are 'screening' and 'assessment of risks'.

Screening. The statutory guidance on inspection (set out in Chapter B) introduces a requirement that each local authority should adopt a "strategic approach to the identification of land which merits detailed individual inspection".[22] The terms in which the statutory guidance amplifies this requirement make it clear that this strategic approach is to be risk-driven, and related to the specific circumstances and history of the individual local authority's area.

In particular, the approach is required to:

"(a) be rational, ordered and efficient;
(b) be proportionate to the seriousness of any actual or potential risk;
(c) seek to ensure that the most pressing and serious problems are located first;
(d) ensure that resources are concentrated on investigating in areas where the authority is most likely to identify contaminated land; and

---

[22] Paragraph B.9, Circular 2/2000, Annex 3.

(e) ensure that the local authority efficiently identifies requirements for the detailed inspection of particular areas of land".

It is clear that under this an arbitrary approach that, for example, singles out particular past land uses for detailed inspection without considering other aspects of potential risks, would not be acceptable.

Assessment of Risks.   The next stage in inspecting individual sites continues the risk theme, in that it is all about identifying particular contaminants that present and are capable of causing harm or the pollution of controlled waters and assessing the probability of their doing so—in the terminology of Part IIA, identifying any 'pollutant linkages'.

The risk assessment process on a site is then completed by an evaluation of the degree of risk presented by any identified pollutant linkage. Is significant harm being caused by the pollutant linkage, or is there a significant possibility of such harm being caused? Or is the pollution of controlled waters being caused, or likely to be caused? In any case where this applies, the land is determined to be 'contaminated land'; the pollutant linkage is subsequently described as a 'significant pollutant linkage'.

Risk Management

Remediation versus Clean-up.   Identification of any contaminated land triggers the next duty to secure 'remediation' of the land. The definition of remediation in Part IIA makes clear that this, too, is a risk-based concept.

The core of this statutory definition of remediation is the part described in the statutory guidance as "remedial treatment action". (For drafting reasons, remediation as defined also encompasses detailed site characterisation—described as 'assessment action'—and post-remediation monitoring—'monitoring action'.) A remedial treatment action is defined as:

"the doing of any works, the carrying out of any operations or the taking of any steps in relation to any . . . land or waters for the purpose

(i) of preventing or minimising, or remedying or mitigating the effects of, any significant harm, or any pollution of controlled waters, by reason of which the contaminated land is such land; or
(ii) of restoring the land or waters to their former state".[23]

The first part of this definition is clearly focussed on the management, control and reduction of risks. The effect of the statutory guidance in Chapter C, which covers the questions of 'what is to be done by way of remediation' and the 'standard to which any land is . . . to be remediated', stresses this risk management approach. This is particular the case in the parts of the guidance dealing with the standard of remediation.

---

[23] Section 78A(7)(b), EPA 1990.

The second part of this definition, taken at face value, enables a 'clean-up' approach to be adopted that focusses solely on dealing with the contaminant. However, this option is closed down, except in specific circumstances, by the statutory guidance in Chapter C. The provision was included largely for the avoidance of any potential doubt as to the extent of the remediation requirements of the Part IIA regime. For example, a similar 'restoration to its former state' definition is used under water legislation to require the re-stocking of watercourses with fish after pollution incidents; the drafters of Part IIA did not wish to find that this kind of requirement could not be imposed, should it be necessary.

Options Appraisal and the Standard of Remediation.   In a regime driven only by arbitrary exceedance of limit values for particular contaminants, the question of 'standards' for remediation would be conceptually quite simple. They could take the form of numerical target values for the same contaminants, set either at or below the triggering limit values. Leaving aside the difficulties of setting comprehensive limit values for all substances in all circumstances, what this approach also inevitably implies is that the only permissible approaches to remediation would be those that acted directly on the contaminants themselves.

By contrast, Part IIA builds in a more flexible approach reflecting the wider concept of 'managing' the risk caused by a contaminant, taking other issues into account at a site by site level—and it does not prejudge how that is achieved. Whilst this does, potentially, open the field to a much wider range of approaches to remediation, it of course makes the process of defining remediation standards rather more complex.

The approach taken in the Part IIA guidance on remediation is, initially, to consider separately each identified significant pollutant linkage, and to identify for each linkage what action or sequence of actions would form the 'best practicable techniques of remediation' for that linkage. This could take the form of removing or treating the pollutant, breaking or removing the pathway or protecting or removing the receptor. The determination of what would be 'best' in this context involves the consideration of what would be 'reasonable' having regard to the seriousness of the risk and the cost involved, and what would give the best combination of practicability, effectiveness, and durability.

The 'standard' of remediation that needs to be achieved on a particular site is therefore whatever degree of risk management would be achieved by the use of the 'best practicable techniques' for remediating each of the significant pollutant linkages on the site.

This standard can then be used to evaluate any alternative approaches to remediation. These might, for example, be remediation approaches that sought to manage the risks for more than one identified significant pollutant linkage simultaneously.

Monitoring.   As suggested above, the Part IIA enforcement process can require action to be taken to 'monitor' sites after remedial treatment action has been carried out. This obligation covers significant pollutant linkages that have already been identified, and is intended to pick up any changes that may occur in the condition of their component pollutants, pathways or receptors.

Over and above this, there are three further levels of monitoring and review:

- the local authority retains its duty to 'cause its area to be inspected from time to time'; this is not conceived as a one-off process, and it might be relevant for identifying any new significant pollutant linkages;
- more broadly, the local authority is required to 'keep its [inspection] strategy under periodic review';[24] new priorities might be identified, for example, in the light of experience gained in earlier phases of inspection and the processing of problem sites through the remediation process; and
- the Environment Agency has a duty to 'prepare and publish a report on the state of contaminated land',[25] building on information generated by the operation of the Part IIA system; this information will then potentially feed into wider processes of policy evaluation and review.

5 Making People Pay

The Importance of 'Liabilities'

Questions of liabilities were critical to the development of Part IIA, not only to ensure that the resulting regime delivered an effective enforcement mechanism for regulators to use, but also to reassure stakeholders that any liabilities they might face would be reasonable, fair and predictable.

The existing statutory regimes dealing with land contamination problems had a variety of different liability tests, each of which presented there own difficulties in practice when applied to land contamination, and were not integrated in their approaches:

- *Statutory nuisance.* Under the statutory nuisance regime in Part III of the Environmental Protection Act 1990, liability fell in the first instance on 'the person responsible for the nuisance', this term being expanded to mean 'the person to whose act, default or sufferance the nuisance is attributable'.[26] Responsibility would pass to the owner or occupier of the premises if no such person could be found or if 'the nuisance arises from any defect of a structural nature'.[27]
- *Anti-pollution works.* Under section 161 of the Water Resources Act 1991, the National Rivers Authority (and then the Environment Agency) could take action to prevent polluting matter entering controlled waters, and could recover its reasonable costs from the person who "caused or knowingly permitted the matter in question to be present at the place from which it was likely . . . to enter controlled waters".[28] The practicability of this mechanism was hampered by the 'act first, recover later' structure; this was corrected by the creation of 'anti-pollution works notices' under section 161A.[29]

---

[24] Paragraph B.13, Circular 2/2000, Annex 3.
[25] Section 78U, EPA 1990.
[26] Section 79(7), EPA 1990.
[27] Section 80(2), EPA 1990.
[28] Section 161(3)(a), Water Resources Act 1991.
[29] This new section was inserted by paragraph 162 of Schedule 22 of the Environment Act 1995.

- *Closed landfills.* Section 61 of the Environmental Protection Act 1990 would have given waste regulation authorities a duty carry out works to prevent harm to human health or the environment from closed landfill sites, recovering their reasonable costs from 'the person who is for the time being the owner of the land'.[30] The draconian nature of this liability provision, which was to be limited only by the consideration of hardship, coupled with a definition of a 'closed landfill' that might in practice have caught up almost all man-made land contamination, meant that this provision was never brought into force, and was repealed in 1995.

In addition to the difficulties facing regulators, wider stakeholders had become sensitised to the question of liabilities through the early 1990s in particular. In part, this reflected a new realisation on the part of many stakeholders that land contamination was something for which there might be liabilities, with fears about consequential effects on property values. These concerns were exacerbated by a general slump in the property market. The fears were also given added focus by comparison with international experience, particularly in the United States under the Superfund programme—this was perceived as imposing high liabilities, on a joint and several basis, across a very wide field of 'potentially responsible parties'.

The effect of these concerns about liabilities had wide potential impacts. Fears about 'residual liabilities' were acting as a significant break on private sector investment in brownfield redevelopment projects. There were also threatening severe dislocations in the property market more generally. A further issue was the potential impact on credit-worthiness of small businesses, which might have bank overdrafts and other loans secured against land assets whose value was potentially compromised.

All of these factors taken together created a need, in the development of the Part IIA legislation and guidance, to go a long way to answer hypothetical questions about who might be liable to pay for remediation in a wide range of different circumstances. Without these pressures, it might have been possible to rely on a more broad-brush approach, for example simply requiring regulators to allocate liabilities amongst liable persons in such manner as they considered fair and reasonable in the circumstances for each site. However, the particular circumstances surrounding the introduction of Part IIA required this allocation to be more defined.

'Polluters' and Part IIA

A core policy requirement for the development of the Part IIA regime was that it should reflect the 'polluter pays principle'. This is internationally accepted as the basic paradigm for environmental liabilities and responsibilities. However, implementing the polluter pays principle for historic contamination is not a simple task. There may, for example, have been a considerable 'chain of custody' for different contaminants over the time they have been in the ground.

[30] Section 61(7), EPA 1990 (never implemented, and repealed by the Environment Act 1995).

Furthermore, if the focus of the regime is on risk—inevitably involving the consideration of pathways and receptors as well as the contaminant—is it really appropriate for liabilities to fall solely on someone responsible for the presence of the contaminant, as would seem to be implied by a common-sense interpretation of the term 'polluter'?

Part IIA seeks to deal with the issues, firstly by the way the concept of a 'polluter' is translated into a definition of a liable person in the primary legislation, and secondly through the statutory guidance on allocating liabilities between different liable persons.

The main liability for any remediation action falls on any person who has "caused or knowingly permitted the substance . . . by reason of which the contaminated land in question is such land to be in, on or under that land".[31] DETR has stated that it interprets this test to relate to the *presence* of a substance in the land, and not just to the process by which that substance came to be in the land.[32] On this basis it would encompass the sense of knowingly permitting the continued presence of a substance in the land, picking up the 'chain of custody' idea outlined above.

The statutory guidance in Chapter D, which deals with circumstances where there is more than one 'appropriate person'—as those who are liable for remediation are termed in Part IIA—for any particular significant pollutant linkage. The guidance sets six 'exclusion tests', that is grounds under which one or more of the appropriate persons should be treated as not being liable, and then provides cascaded series of 'apportionment' rules for dividing liabilities as fairly as possible amongst those that remain.

The six exclusion tests in particular are intended to focus liability on those who are most responsible for the 'risk', rather than just for the fact of the presence of a contaminant:

- *Test 1*—'excluded activities'—has the broad effect of removing from liability those who have been caught in the net only by the actions or inactions of others who should shoulder the full responsibility.
- *Tests 2 and 3*—"payments made for remediation" and 'sold with information'—deal with circumstances where one appropriate person has, in effect, already met his obligations either by making an explicit payment to another liable person or by selling the land, potentially at a reduced price.
- *Tests 4, 5 and 6*—'changes to substances', 'escaped substances' and 'introduction of pathways or receptors'—deal with different types of circumstances where one appropriate person has, in effect, left the site in a safe condition (notwithstanding the fact that he has caused or knowingly permitted a contaminant to be present) only for another person subsequently to change the condition of the land such that there is now a significant pollutant linkage. The tests were included in effect to transfer liability to the person who made the difference.

---

[31] Section 78F(2), EPA 1990. The following sub-section clarifies that any such person becomes liable only for remediation actions which are referable to the particular substance or substances for which that person is responsible; it is not a system of joint and several liability for the site as a whole.

[32] Paragraph 9.8, Circular 2/2000, Annex 2.

Owners (and Occupiers)

Part IIA also has to deal with circumstances where no polluter can be found to pay for particular remediation actions. These might be quite common, given that the regime is intended to deal with contamination from the past—it may not be possible, from a legal or technical perspective, to identify who is responsible for the contamination, or it may be well known but the responsible person or company may have long-since ceased to exist.

In these circumstances, liability under Part IIA passes to the 'owner or occupier for the time being of the contaminated land in question'[33]—the *current* owner or occupier. However, this passage of liability does not take place where the significant pollutant linkage involved has controlled waters as its receptor; in a case of that kind, the liability becomes 'orphan' and responsibility for funding remediation passes to the taxpayer.

The whole question of liabilities falling on 'innocent owners' was—and remains—a hotly debated political topic. The approach taken by the Government was to follow the broad pattern of existing potential liabilities; in particular, the statutory nuisance system has a very similar approach of liabilities passing to the current owner where "the person responsible for the nuisance cannot be found".[34]

The definition of 'owner' adopted in Part IIA[35] is complicated, and reflects the progressive development of equivalent definitions in other regimes to prevent evasion and attempts to hide information about the ownership of land. For the same reason, the 'occupier' is also potentially included within the net of liability—this is to deal with cases where the ownership of the land has been hidden.

Chapter D of the statutory guidance provides a framework for allocating liabilities in cases where there are two or more persons falling within the heading of 'owner or occupier'—there could, for example, be a linked series of leases over the same land, or the 'contaminated land' could encompass more than one ownership plot. The guiding principle in the guidance is that, wherever possible, the allocation of liabilities should reflect the relative value of the capital interests in the land held by the various owners and occupiers.

Dealing with Complex Sites

The basic role of the statutory guidance in Chapter D is to deal with complexities in the allocation and apportionment of liabilities that are beyond the scope of the relatively simple basic rules set out in the primary legislation. The discussion above has outlined the way that the guidance deals with cases where there is more than one 'appropriate person' who might pay for the remediation that is referable to a single significant pollutant linkage. The statutory guidance also deals with cases where there is more than one significant pollutant linkage—multiply contaminated sites with potentially different people being responsible in different ways for different contaminants.

The approach adopted is believed to be unique, in international terms, and is

[33] Sections 78F(4) & (5), EPA 1990.
[34] Section 80(2)(c), EPA 1990.
[35] Section 78A(9), EPA 1990.

built up from the underlying 'significant pollutant linkage' concept that runs through the Part IIA system. What it does is, in effect, to start off by modelling the site as if it were a collection of different sites, on each of which there was only a single significant pollutant linkage.

The enforcing authority first considers what remediation would be carried out on each single significant pollutant linkages (which is something it has to do, in any event, in order to determine the appropriate 'standard of remediation'); the authority then works out who would pay for this, and how much this would cost. Next, the authority considers the site as a whole, and what can be done as a combined 'remediation scheme'. Some individual remediation actions may apply only to individual pollutant linkages; in such a case, it is clear that those liable for those pollutant linkages pay for those remediation actions. But where remediation actions apply to two or more pollutant linkages, the costs are apportioned between the relevant liability groups in proportion to the costs they would have faced if theirs were the only pollutant linkages.

This approach provides a methodology for dividing costs on a fair basis in potentially very complex circumstances. By directly using the costs of dealing with each linkage as the basis for apportionment it avoids any need for the authority to make judgements about the relative seriousness of different contamination problems on the site. It is intended that this will introduce a degree of predictability and objectivity in the way the apportionment works.

6   Part IIA as a 'Regulatory System'

Enforcement versus Voluntary Action

Although Part IIA is essentially an 'enforcement' regime, it is intended to play a part in encouraging the voluntary remediation of land affected by contamination. For example, the discussion above of the Government's objectives in introducing Part IIA noted the role in this process of reducing uncertainties about 'residual liabilities' on brownfield development sites.

DETR also hopes that the existence of a clear regulatory framework will encourage companies to take their own action to identify and remediate problem sites for which they are responsible, pre-empting any formal intervention by regulators. In April 2000, Michael Meacher MP, the Minister of State for the Environment, wrote to the chairmen of the top UK-registered companies on this specific issue, challenging them to develop their own company-specific strategies for dealing with contaminated land and, critically, to allocate the necessary financial resources to pay for investigation and remediation work. He noted that the regime had been made as 'business friendly' as possible; it was now time for businesses to reciprocate and to 'act responsibly, where they are responsible'.

Even within the Part IIA system there are opportunities for voluntary action to replace formal enforcement action. There is, in particular, a requirement for enforcing authorities to consult those who will have to pay for remediation before serving any remediation notice. This consultation process is intended not only to enable technical issues to be resolved before there is any need to invoke formal appeals procedures, but also to give landowners and others the opportunity to bring forward their own suggestions for how best to remediate the land. Many companies, for example, regard being subject to enforcement action as a failure of their overall corporate approach to environmental issues, and will comply with the necessary requirements without enforcement being needed. In other cases, landowners may wish to bring forward wider schemes to remediate the land, for example for redevelopment; this might, for example, enable them to fund the remediation itself out of the proceeds of an increased land value.

In any case where voluntary remediation action is taking place on land that has been formally determined to be 'contaminated land', those carrying out the remediation are required to prepare and publish a 'remediation statement' describing the work. This is then included on the enforcing authority's register, not only to make sure that the information contained on the register is as comprehensive as possible, but in particular to ensure that there is transparency and accountability in the way this part of the regime operates.

Local Authorities as Lead Regulators

The main regulatory roles under Part IIA fall to local authorities. As the Environment Act 1995 which introduced Part IIA also created the Environment Agencies, there was clearly an opportunity to give this role to the these new Agencies. However, a deliberate decision was taken to give the main role to local authorities. This reflected the following factors:

- Local authorities already had equivalent duties, for example under the statutory nuisance system, to deal both with land contamination problems and other local environmental quality issues. In many cases, individual authorities had already made significant progress in identifying and dealing with problem sites in their areas, using a mixture of regulatory powers and funding provided by the Government through the Contaminated Land Supplementary Credit Approval Programme;
- Keeping regulatory functions for contaminated land with local authorities would give the opportunity for the development of significant practical and technical synergies with other local authority functions that deal with land contamination, in particular planning and development control and economic regeneration; and
- Local authorities are well used, both at corporate level and at the level of their individual officers, to working with local communities. This is likely to be centrally important in 'risk communication' relating to contamination risks on housing and school land, in particular.

Additional funding has been provided to local authorities to reflect their new duties under the Part IIA regime. A total of £12 million per annum has been added to totals for 'standard spending assessments' for revenue expenditure; this is then reflected in the provision of Revenue Support Grant. Actual decisions on allocations of revenue budgets between different service sectors are the responsibility of individual local authorities—the additional funding is not ring-fenced.

In addition, the long-standing supplementary credit approval programme for contaminated land assessment and remediation has been expanded to help meet capital expenditure needs of authorities in dealing with individual sites. Total provision in this programme is now running at £21 million per year.

The Roles of the Environment Agency. In addition to having a central role in the production of technical best practice and in providing information and guidance to local authorities on particular aspects of the regime, the Environment Agency has been given the task of acting as 'enforcing authority' for defined categories of contaminated land, described as 'special sites'. In effect, these are all sites where the synergy argument runs in favour of the Environment Agency having the lead role, either because of the nature of the risks on the site (for example, major water pollution cases) or because the Agency is already likely to be present 'on site' as a regulator under other pollution control regimes.

Part IIA as 'Law'—the Role of Guidance

In comparison with many environmental protection regimes, a large proportion of the detail of the Part IIA regime as a whole is set out in the primary legislation, rather than in secondary legislation. The basic structure of the Part IIA system mirrors the existing statutory nuisance provisions which it was replacing for land contamination problems. The outcome of the Government's policy review was also clear that the regime covered many issues that needed primary Parliamentary debate and scrutiny.

Where the Part IIA system also stands out from other environmental regimes is in its reliance on statutory guidance and, in particular, in its use of guidance with which authorities are required to 'act in accordance'. As a legal tool, statutory guidance has been described as providing a structure for the exercise of statutory discretion.

Usually, the authorities exercising this discretion are required to 'have regard to' formal guidance, or to 'take it into account'. In these circumstances this formal guidance cannot simply be ignored—a decision taken that was taken without taking into account the stipulations of any relevant statutory guidance would be open to ready legal challenge. But there may be a wide range of factors, perhaps not discussed in detail or directly addressed in the statutory guidance, in individual cases. The way in which the authority takes the guidance into account has therefore to some extent to be flexible.

The Part IIA regime has two specific instances of this kind of statutory guidance: enforcing authorities are required to 'have regard to' guidance (a) on what remediation is to be carried out on an individual site, and (b) on whether their full liability to pay for remediation should be imposed on any individual or company. But Part IIA is innovative in creating a new category of statutory guidance—guidance with which authorities are required to 'act in accordance'. This guidance is intended to be more prescriptive than the 'have regard to' guidance.

This type of guidance is central to the regime, as it is used to shape the basic definition of 'contaminated land' itself, to set out how local authorities should go about finding any such land, and to determine how liabilities should be divided up in cases where there are multiple liable parties according to the basic provisions of the legislation. These were all questions of importance during the passage of the Part IIA legislation through Parliament, and are clearly of vital significance to the impact of the regime in practice. This importance was also recognised in the requirement for drafts of the guidance to be subject to Parliamentary approval before they were formally issued.[36]

Although this guidance still adopts the typical verbal form for guidance of setting out what authorities 'should' do, the element of compulsion implied by the 'act in accordance' requirement is reinforced by the wording of possible grounds of appeal against any remediation notice. For example, Regulation 7(1)(a) provides for an appeal against a remediation notice on the grounds

"that, in determining whether any land to which the notice relates appears to be contaminated land, the local authority . . . *failed to act in accordance with guidance* issued by the Secretary of State" (emphasis added).

The Role of 'Technical' Advice and Guidance

At each of the key stages of the process, enforcing authorities need to address particular technical questions. Some of these are introduced in the statutory guidance, but more detailed technical judgement will be needed to support and deliver the overall requirements of the regime and its different stages.

In parallel to the work on developing the Part IIA regime, the then Department of the Environment worked on the development of technical guidance. This programme covered both the overall risk assessment and risk management sequence for land contamination management generally (which is also followed by the structure of Part IIA) and specific applications in identifying and dealing with problem sites under Part IIA. This applied in particular to guidance on risk assessment.

Responsibility for management of this programme passed to the Environment Agency in late 1996. Some of the detail is discussed elsewhere in this volume and further information can be found through both the DETR and the Environment Agency web sites. It should be noted that given the complex nature of land contamination and the wide range of guidance being produced it is not the intention to produce technical advice and guidance with a formal legal standing within the overall regime, although regulatory authorities can be expected to have their own quality assurance procedures in place to ensure that they are using appropriate and defensible technical approaches which are relevant to Part IIA.

7 The State of Play

Implementation and Devolution

The primary legislation for Part IIA covers 'Great Britain'—that is England, Scotland and Wales. However, following devolution implementation is handled separately in each of the three countries, with separate regulations and statutory guidance being issued. In England, Part IIA came into force in April 2000. In

---

[36] Section 78YA, EPA 1990.

Scotland, this happened in June 2000. In Wales implementation is not likely to take place until the Summer of 2001.

Part IIA itself will not apply in Northern Ireland, although it is planned that a regime with a very similar structure and approach will be introduced under the terms of the Waste and Contaminated Land Order 1997.

Progress on the Ground

At the time of writing, English local authorities have four months left in which to prepare, consult on and publish their 'inspection strategies'—the statutory guidance says that this should be done by July 2001. Information on around 90% of local authorities from the Environment Agency, which is seeking to assist individual local authorities in this process, suggests that most authorities are making good progress to meet this target. Information management is a key component of the local authorities work, and it is clear that many are considering how best to integrate the information which they will collect as part of their Part IIA duties with their other roles and systems.

Site-level regulatory activity has also started. Again reflecting information available to the Environment Agency at the time of writing, twelve sites have been formally determined to be 'contaminated land' under the Part IIA regime. Active discussions and investigations are continuing on another 50 sites.

Officers from many authorities participate in local and regional groupings of authorities, to identify opportunities for pooling experience and jointly developing expertise. To support this, the Environment Agency has been developing further advice notes, including a guide to the regime, and a training package, on behalf of DETR and the Local Government Association to provide a detailed basis for authorities to develop their skills and implement the regime effectively.

8 What Next?

Other Initiatives

The implementation of Part IIA, even once it is complete across the UK, will not represent the end of policy development on land contamination issues. For example, the following areas of development are already in train:

- The publication in March 2001 of a draft 'soil strategy'.[37]
- The Pollution Prevention and Control regime,[38] which is coming into force over a seven year period starting from 2000, improves the controls over new soil contamination from industrial processes.
- The introduction of a specific regulatory regime for the control of land remediation activities and processes is under consideration by DETR and the Environment Agency.
- DETR is developing a new Planning Policy Guidance note (PPG) dealing specifically with development on land affected by contamination. This will

[37] *The Draft Soil Strategy for England—a Consultation Paper*, DETR and MAFF, 2001.
[38] Pollution Prevention and Control Regulations 2000.

replace and amplify the existing guidance set out in PPG 23.[39]

● The Government confirmed in its Budget Report on 7 March 2001 that it would be introducing a new accelerated tax relief for the costs of contaminated land remediation in the context of redevelopment projects.

The Future?

Contaminated land has always been described as a complex problem, in terms of policy development, in the range of stakeholders involved and not least in the technical and scientific aspects. Part IIA reflects that complexity, and as a new regime it of course presents a challenge to those in the front line now, as well as to their advisors and wider 'customers', to become familiar with it and implement it effectively. However, the substantial development of understanding, partnership solutions and the overall policy and regulatory tools over the last decade should now start to make a real and positive difference in dealing with this industrial legacy and creating a better environment in the future.

[39] PPG23 *Planning and Pollution Control*, DOE, 1994.

Identifying and Dealing with Contaminated Land

MARK KIBBLEWHITE

1 Introduction

Land is a finite and valuable resource. Identifying and dealing with contaminated land is important in order to support increased quality of life for communities and conservation of biodiversity.

Contamination of land can harm humans and the environment and needs to be controlled. Uncertainty about the existence and potentially hazardous nature of land contamination reduces confidence in existing and planned development on brown field sites and so increases demand for green field sites.[1]

During recent decades, considerable areas of contaminated land have been identified and many of these have been dealt with, so that they can be used safely for new development or returned to a semi-natural condition. However, much contaminated land is either not yet identified or is described inadequately. Furthermore, knowledge is incomplete about the impacts of land contamination on humans and natural systems.

Although the majority of contaminated land is owned privately, a range of stakeholders usually have an interest in its management. The needs and expectations of these stakeholders are likely to be different and are often conflicting. Effective communication between them is critical. The owner of the land will wish to ensure that its value is optimised and that liabilities to third parties are known and where possible contained. Developers[2] require sound information on the extent and importance of contamination, so that the costs of dealing with it to allow safe use can be assessed accurately. Regulators are charged with ensuring that the public interest is protected, including harm to human health, the wider environment and economic activities. Those who live or work on contaminated land or close to it have a direct interest in ensuring that they are not being harmed.

The acceptability of the risks presented by contaminated land sites is not a

---

[1] Urban Task Force, *Towards an Urban Renaissance*, SPON, London, 1999.

[2] Environment Agency and the National House Building Council, *Guidance for the Safe Development of Housing on Land Affected by Contamination*, Environment Agency, Bristol, 2000.

Issues in Environmental Science and Technology No. 16
Assessment and Reclamation of Contaminated Land

purely scientific matter,[3] because it involves value judgements. However, science is the foundation for assessment and agreement. A range of scientific and engineering disciplines are relevant, including toxicology, hydrogeology and civil engineering as well as environmental and analytical chemistry. Applying these disciplines to contaminated land in an integrated manner requires effective team working and honesty about the limits of understanding and practices. It is especially important to estimate the uncertainty about scientific conclusions and communicate this to all the stakeholders.

In this chapter, three aspects of contaminated land management are considered. Firstly, the identification of contaminated land sites is discussed. Secondly, the process of site assessment is considered. Thirdly, methods for sampling and testing of contaminated land are outlined.

2   Site Assessment

Background

The purpose of individual site assessment[4] is 'to decide whether or not a site poses actual or potential risks to human health or the environment'. A phased approach is recommended in accordance with recent guidance on environmental risk assessment and management.[5]

The major advantages of this are that it is systematic and it allows decisions to be taken with the appropriate level of detailed information. This is cost efficient, because no more information need be collected and analysed than is necessary to support the required level of decision making.

Three assessment phases are recognised widely. The first is hazard identification: this is described as a Phase 1a investigation. In this, a conceptual model is developed for the site that identifies potential receptors, contaminant sources and pathways, by which contamination can reach receptors. The second phase (Phase 1b) tests the conceptual model, by collecting and analysing quantitative information to support model validation. The final phase is risk assessment (Phase 2) in which the conceptual model is used to estimate and evaluate the actual or potential risks to receptors.

Phase 1a—Hazard Identification

The objective of this phase is to develop a conceptual model for the site, which describes the important source–pathway–receptor linkages for the site and makes a preliminary assessment of their importance.

Figure 1 shows a flow diagram for a Phase 1a risk assessment.

---

[3] Scotland and Northern Ireland Forum for Environmental Research, *Communicating Understanding of Contaminated Land Risks*, SNIFFER Project Number SR97 (11) F, SNIFFER, Stirling, 1999.

[4] Department of the Environment, Transport and the Regions and the Environment Agency, *Model Procedures for the Management of Contaminated Land*, CLR 11, Department of the Environment, Transport and the Regions, London, in press.

[5] Department of the Environment, Transport and the Regions, Environment Agency, Institute for Environment and Health, *Guidelines for Environmental Risk Assessment and Management*, The Stationery Office, London, 2000.

Figure 1 Flow diagram for
Phase 1a risk assessment

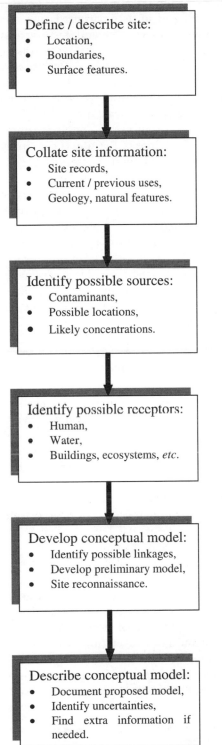

Once the boundaries of the site are defined, a physical description can be prepared of the site itself and of its environs. Using public, commercial and private records, the important current and historic features of the site are identified. These include boundaries, buildings, waste disposal areas and made-ground, topography and natural features—such as surface waters, underlying geology and hydrogeology, and important habitats. Usually, sets of maps and photographs are used to collate and present this information; however, geographic information systems are used increasingly.

The possible presence of sources of contamination is then assessed by reviewing the current and past use of the site. If the site is in use, then possible sources of contamination may remain. These are identified, by reviewing the current storage, handling, transport and use of potentially hazardous materials. Examples of possible sources of contamination are leakage of hydrocarbons from underground tanks and piping, drum storage areas or a process from which fugitive dust emissions are likely.

Whether the site is in use or not, historic sources of contamination have to be considered and often are the main concern. Information on possible sources of historic contamination sources can be found in company records, local authority and regulators' archives, historical maps and old photographs. Much of this information can be available from commercial data houses, regulators and others, often in electronic format. Industry profiles have been published[6] that assist in identifying what types of contaminants are associated with particular industries.

Information that has been assembled on potential sources is analysed to identify the types and locations of possible contamination on the site. These are then mapped on to the physical description of the site and an initial identification is made of receptors that might be at risk. Consideration of the possible pathways between potential sources and receptors leads to a preliminary conceptual model for the site, in terms of candidate source–pathway–receptor linkages. The preliminary conceptual model is then tested and refined by observation, during a site reconnaissance.

Before making a site reconnaissance, health and safety concerns need to be identified and addressed, by preparation of a plan for safe access and personal protection. Sites often present physical dangers, such as deep water and insecure banks, as well as possible exposure to hazardous chemicals.

The objective of site reconnaissance is to explore the assumptions and conclusions of preliminary work that has been based on paper and other records. It should include land adjacent to the study site, where access is available. Firstly, the physical site description is confirmed as accurate. Then it can be extended by observations, such as the presence of unrecorded features, tanks, waste tips, surface water pounding and flow, vegetation and surface types and conditions. Secondly, the general presence and extent of sources can be confirmed. Thirdly, the presence of possible receptors can be confirmed or eliminated. Fourthly, pathways by which contamination can reach receptors, such as drains, streams or surface run-off are identified, although obviously sub-surface and airborne

---

[6] Department of the Environment, *Documentary Research on Industrial Sites*, CLR 3, Department of the Environment, London, 1994.

transport is less easily observed. Fifthly, careful observation may provide evidence of actual harm to receptors, such as pollution of surface waters or damage to vegetation.

Keen visual observation is the most important task during a site reconnaissance. However, some basic in-field chemical testing equipment is useful. Commercial kits are available, mainly based on colorimetric reactions. These kits can be used to check for the presence or absence of common contaminants and to measure soil pH. However, the results obtained using these test kits should only be regarded as qualitative, not least because samples taken at the surface may not represent the bulk of materials on the site. For site reconnaissance, the use of in-field instrumental testing is not justified. An exception is hand-held equipment to test for the presence of gaseous and volatile contamination—photo-ionisation and other detectors are useful where such contaminants are anticipated. While a Phase 1a study does not normally include intrusive investigation, a small number of samples may be taken for laboratory analysis. These can be important when waste and other potentially contaminated materials of unknown composition have been spread, deposited or are stored on the site.

Following the site reconnaissance and analysis of the information that it provides, conclusions are made and reported on the relevant pollutant linkages that exist or are likely to exist in relation to the site. This requires systematic description of the receptors that are exposed, the sources of exposure and the pathway by which this occurs, together with an initial assessment of likely hazard—in terms of actual, probable or possible harm that is expected or has been identified. Where it is concluded that possible linkages are absent, it is important to present all the supporting evidence for this conclusion. If actual harm to a receptor is identified, then recommendations are given for urgent remedial works, including more detailed investigation, if needed.

Phase 1b—Risk Assessment

The objective of this phase is to test the conceptual model developed in the Phase 1a assessment, to quantify the extent of sources and the possible transport of contaminants to receptors, and to classify confirmed pollutant linkages in terms of the risk presented to receptors.

Figure 2 shows a flow diagram for a Phase 1b risk assessment. This phase often requires a site investigation, including the collection of samples from trial pits or boreholes. A detailed intrusive site investigation is less likely, as this may be more effectively carried out as part of a Phase 2 risk estimation. Whatever, the first stage within Phase 1b is collation and analysis of all existing information to provide the best description of the proposed conceptual model of the site. In particular, the collation of all existing site records and plans provides a basis for the best possible understanding of the site, including sub-surface features and conditions, so that further site work is targeted effectively.

A first simplifying step is to break the site into a number of smaller zones that are expected to contain different types and levels of contaminants. Typically, these zones correspond to areas of the site on which different current or past activities have taken place. For example, on a former manufacturing site those

Figure 2 Flow diagram for
Phase 1b risk assessment

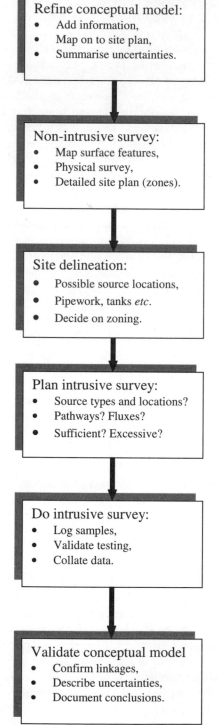

Refine conceptual model:
- Add information,
- Map on to site plan,
- Summarise uncertainties.

Non-intrusive survey:
- Map surface features,
- Physical survey,
- Detailed site plan (zones).

Site delineation:
- Possible source locations,
- Pipework, tanks *etc*.
- Decide on zoning.

Plan intrusive survey:
- Source types and locations?
- Pathways? Fluxes?
- Sufficient? Excessive?

Do intrusive survey:
- Log samples,
- Validate testing,
- Collate data.

Validate conceptual model
- Confirm linkages,
- Describe uncertainties,
- Document conclusions.

areas where metal-plating operations took place can be zoned separately from those used for offices. Information on the location and extent of services, such as pipe work and surface drainage, is especially valuable as these identify areas where past spillage and leakage may have led to ground contamination. This process of site zoning may lead to a number of overlapping areas, with areas of the site falling into more than one zone when land-use has changed or land has supported more than one activity.

A more complete knowledge of the physical layout of the site and its below ground features can be obtained by non-intrusive survey methods. Ground-penetrating radar is able to locate buried features such as pipes and tanks. Soil conductivity may reveal buried objects and areas of made-ground or waste disposal. From the results of these types of survey, areas for possible intrusive investigation are identified and the zoning of the site can be refined.

Careful planning and design of intrusive investigation is always needed. This is critical to ensure that the investigation is sufficient, in terms of both the quantity and the quality of information collected, while not going beyond requirements. Site investigation is only a tool for achieving the objectives of the Phase 1b assessment: it is not the assessment itself.

Firstly, sampling and testing should be chosen that addresses the specific hypotheses contained in the proposed conceptual model. The objectives of the investigation should link to questions such as: 'Is the existence of assumed sources of contamination proven? Are the assumed pathways between sources of contamination and identified receptors viable? Is the extent and strength of the assumed sources sufficient to drive transfer of contamination to identified receptors?' The investigation is planned around these and other key questions, so that it gives information on the presence or absence of site features, the spatial extent and concentrations of contaminants, the presence of pathways and whether these are active or not, and the dynamics of water and gas movement.

Secondly, the design of the sampling and testing is chosen to provide information that is 'fit for purpose'. To achieve this careful consideration is given to both the sampling strategy and the choice of measurement methods employed in the field or the laboratory (see Section 4). A critical consideration is that contaminated land can only be properly described by considering three dimensions. Variability in composition and properties with depth is as important as that with surface position.

Thirdly, intrusive investigation is carried out carefully, so that further contamination is avoided, for example by damage to buried tanks and pipes or by breaking barriers to movement of contamination, such as clay horizons protecting groundwater.

New information gathered from additional desk studies and any site investigation is then collated and used to provide more complete descriptions of the candidate source–pathway–receptor pollutant linkages identified in the Phase 1a report. Each linkage is considered separately and estimates are made of the extent and strength of the contaminant source, its connectivity to the receptor and the level of hazard that is presented to the receptor. Finally, a preliminary assessment is made of the potential harm that hazards may pose to the receptors. From this analysis it is likely that one or more pollutant linkages will be recognised as the most important in terms of actual or potential risk. An evaluation of the significance of these leads to a decision on the need for more formal risk assessment. For example, if the Phase 1b assessment does not demonstrate the presence of a viable pathway to any significant receptors further assessment would not be justified. On the other hand, if the Phase 1b assessment unequivocally identifies a viable pollutant linkage with a receptor, then a Phase 2 risk assessment is needed.

Successful conclusion of a Phase 1b assessment depends on careful analysis of data to validate the conceptual model and identify those candidate pollutant linkages within each site zone that need to go forward for Phase 2 risk assessment.

Phase 2 Risk Assessment

Overview. The starting point for risk assessment is knowledge of which source–pathway–receptor linkages are plausible for the site that is being assessed. At this stage, the conceptual model for the site should be completed and agreed with key parties, such as regulators.

Where an actual or potential risk from contaminated land is identified, this has to be estimated and its impact evaluated as a basis for risk management. This is the purpose of Phase 2 risk assessments.

Figure 3 shows a flow diagram for a Phase 2 risk assessment. The first stage of Phase 2 assessment is risk estimation; this estimates the potential harm to all receptors within identified source–pathway–receptor linkages. The second stage is risk evaluation; this considers the acceptability of potential harm to receptors, and identifies and assesses preliminary options to deal with unacceptable risks.

For the majority of sites, and certainly for the larger and more complex ones, zones within the site will have been identified that contain different types of source and that have distinct pollutant linkages associated with them. Each of these zones should have an associated conceptual model and be subject to risk assessment.

Risk Estimation. Accepted practice is to use a tiered approach to risk estimation. This has several benefits. Firstly, the scope of the risk estimation exercise can be refined in a progressive and systematic way towards those linkages that are most important and those about which there is most uncertainty. Secondly, new information is collected when it is required, so that information management is cost-efficient. Thirdly, the results of the estimation process can be reported in stages as the overall exercise proceeds, revealing a progressive narrowing of anticipated outcomes that can be fed in to risk management planning.

Tier 1 of the risk estimation process relies on 'generic' risk assessment criteria. These are 'guideline levels' of contamination that can be used to screen out those pollutant linkages that are unlikely to present a significant possibility of significant harm to particular types of receptor. They are derived by using validated models to estimate the quantity of contaminant transferred to a receptor within a generic land use scenario and then comparing the estimates

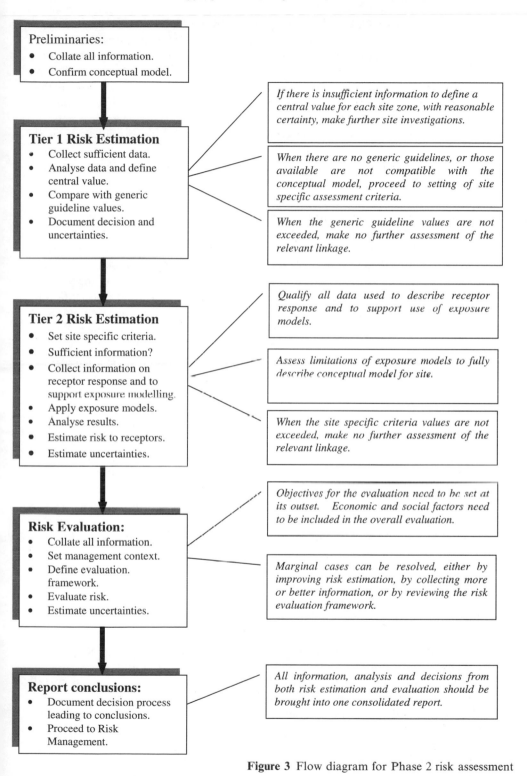

Preliminaries:
- Collate all information.
- Confirm conceptual model.

If there is insufficient information to define a central value for each site zone, with reasonable certainty, make further site investigations.

Tier 1 Risk Estimation
- Collect sufficient data.
- Analyse data and define central value.
- Compare with generic guideline values.
- Document decision and uncertainties.

When there are no generic guidelines, or those available are not compatible with the conceptual model, proceed to setting of site specific assessment criteria.

When the generic guideline values are not exceeded, make no further assessment of the relevant linkage.

Tier 2 Risk Estimation
- Set site specific criteria.
- Sufficient information?
- Collect information on receptor response and to support exposure modelling.
- Apply exposure models.
- Analyse results.
- Estimate risk to receptors.
- Estimate uncertainties.

Qualify all data used to describe receptor response and to support use of exposure models.

Assess limitations of exposure models to fully describe conceptual model for site.

When the site specific criteria values are not exceeded, make no further assessment of the relevant linkage.

Risk Evaluation:
- Collate all information.
- Set management context.
- Define evaluation. framework.
- Evaluate risk.
- Estimate uncertainties.

Objectives for the evaluation need to be set at its outset. Economic and social factors need to be included in the overall evaluation.

Marginal cases can be resolved, either by improving risk estimation, by collecting more or better information, or by reviewing the risk evaluation framework.

Report conclusions:
- Document decision process leading to conclusions.
- Proceed to Risk Management.

All information, analysis and decisions from both risk estimation and evaluation should be brought into one consolidated report.

Figure 3 Flow diagram for Phase 2 risk assessment

obtained with toxicological and other information about receptor responses to contamination. The CARACAS programme made an evaluation of guideline values that are used within Europe.[7] When using these values it is essential to check that the conceptual model on which they are based is consistent with that for the site and that the criteria used to define the values are relevant to the context in which they are to be applied. Guideline values should not be confused with 'Trigger' values (such as the ICRCL values), which set arbitrary levels above which remedial action is required, without reference to a site-specific conceptual model or risk assessment, which approach is not recommended.

If contamination is less than guideline levels, it can be assumed that the risk to the relevant receptor is acceptable and no further estimation or evaluation is required. However, the use of guideline values to confirm that there is an unacceptable risk to receptors is generally wrong, unless the contamination exceeds the guideline values by large margin, certainly an order of magnitude. This is because guideline values are based on a conservative set of assumptions, to allow for a broad range of possible source, pathway and receptor characteristics. If contamination exceeds guideline values then more work will be needed to estimate risks. An initial response should be to check that the data on contamination levels are of sufficient quality to support their accurate comparison with the guideline values. For example, more samples may need to be collected and tested to reduce measurement errors or to better delineate and so reduce the land area within which the source is identified. Depending on the context in which the risk assessment is being made, there may be scope to restrict the options for land use to those for which guideline values indicate there is no unacceptable risk. However, the most likely outcome, when contamination has been found to exceed guideline values, is that site-specific risk estimation is needed. Of course, if no relevant guideline value is available for a receptor–contaminant combination, site-specific risk estimation is unavoidable.

Tier 2 site-specific risk estimation may require additional information, often based on a detailed intrusive site investigation that describes fully the extent and nature of the contamination and the pathways to receptors. The scope for informing the risk estimation by more detailed description of the receptor may be limited, but is also an option.

The Phase 1 risk assessment should have provided a basic understanding of the types of contaminant present on the site and their spatial extent. To support Tier 2 risk estimation, more detailed information is usually needed.

The precise nature of the contaminant on the site may be important to the risk estimation. For example, the Phase 1b risk assessment may have confirmed the presence of polychlorinated biphenyls (PCBs) that could reach humans. However, there is a large variation in the toxicology of individual PCB congeners and so congener-specific measurement is required for risk estimation. Total metal concentrations are sufficient to confirm the presence of a source but metal speciation may influence solubility and so the risk posed to groundwater. The choice of chemical analysis for source characterisation should be informed by

[7] C. Ferguson, D. Darmendrail, K. Freier, B. K. Jensen, J. Jensen, H. Kasamas, A. Urzelai and J. Vegter, (eds.), *Risk Assessment for Contaminated Sites in Europe. Volume 1, Scientific Basis*, LQM Press, Nottingham, 1998.

what may harm receptors and what forms have been assumed in the toxicological data that will support risk estimation.

Intrusive site investigation to provide more detailed knowledge of the spatial extent of contamination should be done in parallel with progressive tiers of risk estimation. Within both tiers of risk estimation, further site and other investigations may be needed as the assessment proceeds. Within these tiers, several more detailed and targeted site investigations may be required. Success depends on a combination of formal sampling design methodology with practical observation and experience. The problem is to avoid missing significant contaminant locations without excessive sampling. Further consideration of these issues is provided in Section 4.

In most cases, pathway characterisation will require a good understanding of water movement on a site if this is included within the conceptual model. This requires knowledge of the site geology, soils and made ground so that the location and extent of water flow can be estimated. Supporting data such as hydraulic conductivity, bulk density and porosity measurements may also be needed. The data requirements of any models that are to be used to estimate risks of water contamination will inform the scope and design of site investigation. The extent of dilution of contamination is critical. Models such as ConSim[8] take account of this dilution.

Characterisation of receptors and their response to contaminants is the third information collection stage within risk estimation that leads to site-specific risk assessment criteria. For humans,[9] generic estimation of soil contaminant uptake, such as that provided by the CLEA[10] model, is subject to quite large safety factors that generally outweigh changes in uptake that might result from site-specific behaviours and exposure that are not taken account of within generic land use scenarios. However, some uptake models allow for contaminant uptake on sites that humans may visit, apart from the particular one being assessed. Depending on the nature and local prevalence of the contaminant, a site-specific allocation may be justified.

For surface and ground-waters there are established contaminant concentration limits and so in general there is little advantage to be gained from attempting more detailed description of receptor response characteristics.

The variations in diversity and function between individual natural ecosystems are too large to allow a broad-brush generic approach to receptor response and some form of site-specific estimation is normally needed. However, the methodology for this is not fully developed at this time. The characteristics of ecosystems that are important to describe and to assess in terms of receptor response are not agreed. Among the proposed approaches[11] is a focus on the contaminant

[8] Environment Agency, *ConSim Contamination Impact on Groundwater: Simulation by Monte Carlo Method,* Environment Agency, Bristol, 1999.

[9] Environment Agency, *Guidance on Site Specific Assessment of Chronic Risks to Human Health from Contamination,* Environment Agency, Bristol, 2001.

[10] Department of the Environment, Transport and the Regions and the Environment Agency, *The Contaminated Land Exposure Assessment Model (CLEA) Technical Basis and Algorithms,* CLR 10, Department of the Environment, Transport and the Regions, London, in press.

[11] Environment Agency, *Biological Stress Indicators of Contaminated Land Ecological Assessment of Contaminated Land Using Earthworm Biomarkers,* Environment Agency, Bristol, 1999.

response of indicator species, such as earthworms, that are judged to have key roles in the ecosystem's functionality. An alternative approach is to assess the response of the ecosystem in terms of energy or carbon transformations or as indicated by reduced diversity, but this requires an understanding of what occurs in the ecosystem when it is in 'good ecological condition', which is not easily described. A pragmatic approach may be applicable where the main concern is protection of specified species, such as when the receptor is a nature conservation site. In this case only individual species' response to contaminants could be estimated but this is easily criticised where that species is dependent on ecological relationships that could be affected by contamination, even if the particular species is not directly at risk. There is an urgent need for research in this area.

The chronic and acute response of agricultural crops and livestock to some contaminants, such as metals, is known from experimentation and guideline values are available. However, site-specific factors such as irrigation of crops and supplementary feeding of livestock can influence receptor exposure and there is a large impact of soil and local environmental conditions on agricultural yields, so field trials may be justified.

Ideally, further characterisation of the source and pathway, combined with understanding of receptor response to contaminants, should lead to a numeric probability of specified harm to the receptor. In reality, risk estimation is often confined to assessing whether threshold values for unacceptable harm are likely to be exceeded, because dynamic relationships between receptor exposure and harm are not generally available. Understanding and reporting the risk factors attached to threshold values is therefore important.

The confidence of risk estimation depends on a combination of the uncertainty associated with any threshold value used and the error of measurements used to test whether that threshold value is exceeded. In most cases, neither have well-defined confidence intervals. Factors that can be used to assist judgements about the robustness of risk estimation include the extent to which safety factors have been included in the derivation of threshold values and the assumed errors of measurement from sampling and testing.

The overall risk estimation should cover all the pollutant linkages identified, broken into site zones as appropriate. Details should be provided of source concentrations, with estimated confidence limits, the receptor characteristics and confirmation of the viability of the pollutant pathway. If generic guideline values are used, then their applicability should be supported. Clearly, full details need to be recorded of any models and data used to generate site-specific values for unacceptable risk, together with a commentary on assumptions that have been made in their application.

Risk Evaluation. The role of the scientist and engineer in the evaluation of risk from contaminated land is a demanding and responsible one. The task is to communicate a complete technical assessment, including the limits of technical certainty. This then informs decision-makers, including non-experts, who have to agree on the future management of the site. All parties will make a better evaluation if there is effective communication of the facts and processes that support and qualify scientific conclusions. If this is not done, then confidence in the overall evaluation will be undermined.

Objectives for the evaluation need to be set at its outset and, if possible, agreed with all those who will make use of, or depend on, its conclusions. However, the scope of the objectives is constrained to those risks identified in previous stages of the site assessment. If others are identified at this stage, or the description of previously identified risks is found to be inadequate, it is necessary to return to the risk estimation stage.

All facts and conclusions relevant to the site need to be considered in the risk evaluation. Information from previous stages in the site assessment is collated and additional facts, particularly future site management details, may assist objective setting and the subsequent evaluation. All of the information available should be collated and ordered around its relevance to the evaluation objectives. For example, if an objective is to evaluate the risk to a groundwater abstraction well, then information that relates to the pollutant linkages to this receptor is collated and ordered.

Objective consideration is needed of the quality of risk estimation. There are many sources of uncertainty in risk estimation. There are errors in the descriptions of the extent and level of site contamination, the transport of contamination to the receptor and the response of the receptor to the contamination. All of these errors require at least qualitative estimation. Then an overall statement of uncertainty of estimation is needed for each risk that is being evaluated.

The general scheme for the technical evaluation is to take each receptor and to describe the level of risk that is presented to it by sources within the site. This leads to a set of key pollutant linkages that drive the overall risk from the site. By reviewing the criteria for identification of unacceptable risk for these key linkages—in terms of the certainty of threshold values, the impact of wrong risk assessment decisions on costs and the quality of site-specific information—a robust risk evaluation framework can be set for identifying unacceptable risk.

Within the risk evaluation framework, the results from the previous risk estimation can be judged as indicating acceptable or non-acceptable risks. However, careful review is needed, to check that the quality of information on which decisions have been made is sufficient.

The results of evaluating risks within the evaluation framework are reviewed in the context of social and economic considerations. The major considerations here are economic benefit relative to the costs of risk reduction and the need to take account of perceived as well as estimated actual risk. Ultimately, the acceptability of risk is not a technical judgement. It has to be determined by agreement around a combination of both technical risk estimation and the broader economic and social consequences of risk that flow from its perception by individuals and the actions that they take in the light of their perceptions. For example, scientific evaluation may show that the risk to children in a communal garden that is contaminated with lead is negligible, when compared to the risk from other sources and when measured against internationally agreed standards. Even so, it may still be concluded that the risk is too high and remedial measures are needed, if action is needed to remove social unease. Ultimately, the decision as to what risk is acceptable has to be agreed by individuals who are at risk, if the risk is to humans or their property, or by informed technical opinion supported by political process, where the risk is to public goods, such as natural systems.

The evaluation is not complete if some risks remain described neither as acceptable or unacceptable. Marginal cases can be resolved, either by improving risk estimation, by collecting more or better information or by reviewing the risk evaluation framework.

Following risk evaluation, a consolidated risk assessment report is prepared, covering the description of the site and the pollutant linkages identified and the overall risk assessment, supported by validated data sets. This sets out unambiguously the main risks posed by the site and whether these are acceptable or not, and may identify preliminary options for dealing with unacceptable risks.

3 Risk Management

Introduction

Risks from contaminated land arise from pollutant linkages that present unacceptable risk to receptors. The purpose of risk management it to make changes that remove these unacceptable risks. This can be done in different ways: the source of contamination may be removed or reduced; the pathway by which contaminants reach the receptor may be broken or attenuated; or the receptor can be removed or modified.

Three inter-related factors are important when choosing and implementing risk management measures: the completeness of pollutant linkage descriptions; confidence in the performance of possible measures; and the effectiveness of management, from design selection, through implementation to any post-completion monitoring. For example, information on pollution linkages available from risk assessment may be sufficient to support using natural attenuation, whereas with less information an *ex situ* soil treatment option may have to be chosen. However, natural attenuation may rely on long term monitoring and if the management is not available to support this, an *ex situ* option may still need to be chosen. Success depends on ensuring that there is an adequacy of information, a treatment option of known and verifiable performance and effective management procedures, supported by audited quality systems.

Risk Management Planning

The starting point for risk management planning is the conclusion from the Phase 2 risk assessment. However, the risk management option that is chosen has to meet management objectives for the site, as well as risk reduction ones. The former will depend on the existing, planned or potential uses of the site. Management objectives and constraints will determine what resources are available for risk management. A combination of desired outcomes for both risk reduction and land management defines the remedial objectives for the site. Typically, these remedial objectives will include a statement of what risks are to be reduced within constraints such as acceptable land uses, costs and timing, and if appropriate, regulatory requirements. Success criteria are needed for each remedial objective to support the evaluation of options. The specification for risk management is critical, as it will inform the final selection of a management option and also form the foundation for verification that risk management has been successful.

Both the risk reduction and the management objectives for the site as a whole may support a set of sub-objectives for zones on the site that have either a particular risk profile or a land management requirement. For larger sites, a number of different remedial objectives may have to be adopted for different site zones, so that the most effective and cost-efficient management option is implemented for the site as whole.

All possible risk management options need to be identified, including different land management scenarios. Then these are screened for their potential to deliver the remedial objectives. For example, if there is a remedial objective to contain costs to within a fixed budget, then a comparison of benchmark costs for different remedial treatments may quickly rule out some more expensive *in situ* source treatment techniques. If there are no constraints on time and there is good infrastructure for long term site management, natural attenuation could be retained as an option. After screening, more detailed information is collected on the risk management options that remain, when further screening may allow elimination of additional options.

Full evaluation of qualified risk management options should be based on selection criteria that have been thoroughly checked, expanded if needed and agreed with the site stakeholders. To allow an overall evaluation, it is advisable to set out a system for scoring and weighting the conclusions for individual and unrelated criteria. Formal cost–benefit appraisal may be required to either separate close contenders or to support initial conclusions.

It is important to obtain as many possible viewpoints on the short-listed options, as this will increase confidence in the decision making process. In this regard, it may be helpful to do one or more pilot studies, depending on the options that are short-listed and the confidence that can be attached to their performance from experience on similar sites. This not only allows assumptions about performance to be tested, but also the collection of site-specific data. Although a pilot may be costly and take valuable time, it may still be justified by the increased confidence it can bring to project costs and timings.

Risk Management Delivery

Risk management actions can extend from more simple measures to the very extensive and complex. In all cases, however, it is essential to follow a formal and documented management system. Without this rigour, the confidence of all stakeholders and the value of the land to society and to its owner may be prejudiced.

When the risk management option involves temporary or permanent engineering works, the management system should provide for design and construction checking, preferably by an independent engineer. As these works usually progress through a series of stages, with a successful outcome from each being critical to successful progression to the following stages, quality control is important. This can include testing to confirm removal of contaminants, the specification of construction materials or the dimensions and performance of structures. The completeness and correctness of this testing should be verified by audit.

Successful delivery is verified against the success criteria within the risk management plan. Several stakeholders may need to confirm this verification. The site owner or developer will wish to know that the proposed management of the site is acceptable and that any new site valuation can be supported. The regulator may have to confirm that residual risks fall below regulatory thresholds. Finally, but most importantly, site occupants need to know that any risks to them have been reduced adequately.

4   Measurement of Site Contamination

Sampling

Generally, contaminated sites are very heterogeneous and this presents difficulties.[12] Simpler sampling strategies that are used for field soil sampling are not effective, as these assume a high degree of site homogeneity. A reliance on statistical theory quickly develops a requirement for very large numbers of samples, which is not realistic in most situations. Usually a pragmatic, stepwise methodology is most effective and efficient, based on a series of increasingly more intensive but spatially focused sampling exercises. The earlier exercises provide information on where contamination definitely exists and where more evidence is required to confirm its absence. Subsequent exercises then focus sampling effort on those areas that remain characterised insufficiently. Confidence limits for the range of contaminant concentrations within each area can be estimated as the exercises progress, so that when the higher limit falls below that of interest, or the lower limit falls above that of interest, no further sampling and testing is made in that area.

On larger sites, a common approach is to sample the whole site on or within a grid, with centres set between 20 and 50 metres apart, and then sample more intensively where higher or anomalous values are reported. This can then be extended by additional sampling at locations where features are noted, such as the presence of waste materials or disturbed ground, or where contaminating activities are known to have taken place. This can be linked to the delineation of site zones for risk assessment purposes.

By subdividing the site into areas with common features that relate to current and past land uses, overall site heterogeneity is broken into areas within which there is more homogeneity. However, within each sub-area, variations in ground conditions are still likely to remain large. Further subdivision of areas can assist and, in particular, areas for sample collection may be usefully delineated by reference to the position of existing or demolished buildings and services. In the end, a judgement is unavoidable as to the type, locations and number of samples that are collected within each area. The basis for this judgement, however subjective, needs to be documented clearly, along with detailed records of sampling locations and procedures.

A 'herring-bone' sampling design has been recommended[13] for areas of relative

---

[12] British Standards Institution, *Code of Practice for Investigation of Potentially Contaminated Sites*, British Standards Institution, London, 2000.

homogeneity. The number of samples taken depends on a combination of the expected heterogeneity of the sampled area and the anticipated difference between the sample mean contaminant concentration and values with which this mean is to be compared. If the purpose of the sampling is to confirm the presence of contamination at high concentrations, then a few samples may be sufficient, but as the concentration level of interest approaches background levels increasing numbers of samples are required to discriminate areas where contamination is present. An iterative process can be followed, by continuing re-sampling until a sample mean is obtained with sufficiently narrow confidence limits to allow its discrimination from, for example, site-specific risk assessment criteria values.

The choice of ground sampling method depends on the nature of the ground and to some extent on access. Hand augers allow sampling to a restricted depth in looser ground. Motorised augers and drills are more versatile but require sufficient access and can be costly (so they may be used more appropriately in Phase 2 investigations). Mechanical excavation of inspection pits has the advantage that the full profile can be examined and sampled with relative ease. Whatever method is used, it is essential to record and retain full sampling logs.

As soon as samples are taken in the field, the concentrations and forms of contamination within them may start to change, so that they are no longer represent field conditions. This applies to both soil and water samples. Procedures for sample collection and transport have to take account of these possible changes. This problem is most severe when measuring volatile contaminants, for which rapid sealing of samples with a minimum of headspace is recommended. Other contaminants that are affected by the changes in redox potential that accompany sample storage, or that may be degraded by microbes, may need to be treated on site to arrest their deterioration.

Non-intrusive Measurement

Field-based non-intrusive methods of contaminant investigation, as distinct from non-intrusive physical investigation, offer several potential advantages. They avoid errors that arise from contaminant changes during sample collection and transport. Often they can be done quickly, so that a larger number of measurements can be made within available resources and time. They lend themselves to a stepwise process, in which sampling density is increased progressively to confirm and identify the extent of areas of higher contaminant concentration. Additionally, non-intrusive methods can be combined with Geo-stationary Positioning Satellite systems and hand-held data processing and storage, so that site data can be readily archived and analysed. To date, however, the application of field-based non-intrusive methods has been limited by the performance of available field instrumentation. One exception is radiometric measurement. Another is X-Ray Fluorescence (XRF) Spectrometry; this can map successfully concentrations of some elements, such as lead, for which XRF is more sensitive. Hand-held photo-ionisation detectors are used widely, but they lack the

---

[13] Department of the Environment, *Sampling Strategies for Contaminated Land*, CLR 4, Department of the Environment, London, 1994.

selectivity needed to provide good quality data on the types and locations of volatile organic compounds, although they can help with decisions about intrusive sampling.

Laboratory Testing

The methods used for laboratory testing of field samples should have validated performance and be conducted within a rigorous quality management system. To avoid reliance on poor data, great care is needed to ensure that the test results reported are derived from methods of known performance that have been applied exactly. This can be confirmed by ensuring that full validation and quality control information is available to support test data.[14]

In most cases, chemical analysis of environmental samples relies on an extraction step, followed by determination of the target chemical species in the extract obtained. Complete extraction is not possible for most contaminants, unless very onerous procedures are followed. As a result, the extent and variability of extraction must be known as otherwise the field concentrations of contaminant may be underestimated. When dealing with water samples it is possible to introduce a recovery spike into the sample, preferably immediately on sampling in the field, in which case the efficiency of the whole testing procedure, including extraction, can be assessed. For soil samples, with the exception of more volatile organic contaminants, dispersion of the spike throughout the sample is very poor, so that this approach does not work.

When chemically aggressive extraction reagents are used on soils, the non-extracted contamination may not be available environmentally. An example is the use of aqua regia to extract metals from soils. This does not remove metals that are strongly bound to silicates and that are unlikely to be mobilised in the field. On the other hand, some methods of extraction are simply inefficient and in these cases account has to be taken of the negative bias that incomplete extraction can introduce into the site data. A particular concern is that most organic contaminants are incompletely extracted from soils by current laboratory techniques. The recent introduction of new and automated techniques, however, has improved extraction efficiency and as importantly reduced its variability. An example is the use of supercritical fluid extraction for polycyclic aromatic hydrocarbons (PAHs). When extracting non-metal inorganic species, extraction efficiency is very dependent on the reagents used and the exact procedure employed, including choice of apparatus, soil mass to extract volume, heating method and contact time.

A variety of instrumental analytical techniques are used to measure contaminants in extracts of both soil and water from contaminated land sites.

A wide range of elements, including arsenic, cadmium, chromium, copper, lead, nickel and zinc can be measured simultaneously with high sensitivity and selectivity using Inductively Coupled Plasma Mass Spectrometry (ICP-MS), a technique which is now replacing the less sensitive and versatile Inductively Coupled Plasma Optical Emission Spectrometry (ICP-OES) for environmental

[14] M. Kibblewhite, L. Heasman and V. Forster, *Development of a validation protocol for methods of analysis of contaminated soil*, in *Contaminated Soil '98*, Thomas Telford, London, 1998, 895–896.

analyses. For the measurement of mercury, Atomic Fluorescence Spectrometry (AFS) after generation of mercury vapour is a more reliable technique than is ICP-MS. Although the latter has a good sensitivity for mercury, background contamination of the instrument is hard to control.

The measurement of non metal inorganic contaminants, such as cyanide and sulfate, is dependent on wet chemistry techniques, albeit usually automated and with final determination using spectroscopy, ion chromatography or ion-selective electrodes.

Information for estimating the risk to receptors from organic compounds requires their specific determination. The measurement of organic contaminants as totals of different chemical species has been a common but undesirable practice. For example, specific congener measurement of polychlorinated biphenyls (PCBs) is now done routinely and is emerging as the standard for PAHs. Typically, analysis of these types of contaminants involves an extraction–concentration step based on sorption or transfer to a different phase, followed by gas chromatography–mass spectrometry (GC–MS) or less commonly liquid chromatography–mass spectrometry (LC–MS). The combination of capillary columns for gas chromatography with sensitive quadrupole based bench top mass spectrometry has proved especially powerful and cost-effective for contaminated land investigations.

Data Reporting

Site investigation data should be reported according to accepted best practice. Full details are needed of sampling locations and depths, associated observations of site conditions and features, the sampling methods employed and the procedures used for sample storage and transport. Details of sample preparation are important to record, as are data on the weights of stones and other materials removed prior to testing. Descriptions of laboratory testing methods, including validation data and quality control information, are essential as these qualify estimates of test bias and precision, without which testing data are incomplete.

For each area or zone of the site that is being considered within risk assessment or management, a representative value for the overall contaminant concentration is needed, with an uncertainty estimate. In many cases, this value is taken as the sample mean. However, the data sets need to be studied carefully to identify how they are distributed, and other estimates of a central value, such as the median, may need to be considered. Estimates of uncertainty may have to rely on conservative judgement as well as on statistical methods. Many data sets are not normally distributed, sample numbers may be low and the overall uncertainty is in any case influenced by a combination of both controlled and uncontrolled sampling and testing errors. For example, whereas the precision of the testing method may be measurable and reported, errors that arise from sample bulking, preparation and subsequent sub-sampling are known less often.

5 Conclusions

The management of contaminated land is challenging because it is characterised by uncertainty. The nature and extent of contamination is often hard to describe adequately. The toxicological or other adverse impact of the contamination on humans and the environment has in many cases to be predicted and rarely can be measured directly. The movement of contaminants from source to receptor through a heterogeneous and ill-defined site matrix is complex to model. However, confidence and effective use of available information and resources can be optimised by taking a phased approach and by ensuring, within each successive phase, that information and actions are of known and documented quality and are communicated completely to all stakeholders.

Contaminated Land and the Link with Health

ANDREW J. KIBBLE AND PATRICK J. SAUNDERS

1 Introduction

Soil plays an integral part in our lives and is inherently linked to public health. For example, many of the essential trace elements which we require in our diet to remain healthy are derived from soils and parent rock material, and low concentrations or the unavailability of these elements in soil can cause dietary deficiencies. Soils can also be contaminated with a range of potentially hazardous substances (both chemical and biological) which, if present at sufficiently elevated levels, can present a potential public health problem. For example, soils may contain elevated levels of heavy metals such as cadmium and lead which can have measurable and often severe effects on local populations. The soils of Cappadocia in central Turkey are naturally rich in fibrous asbestos-like minerals that are thought to be the cause of a rare cancer in local communities[1] while exposure to the bacterium *Clostridium tetani* in soils can cause tetanus. Despite such examples, the effects of contaminated land have, until recently, been relatively ignored and, even today, our understanding of the mechanisms and level of risk associated with contaminated land is poor in relation to air and water.

As we will describe later, investigating the potential health effects associated with contaminated land can be extremely difficult and is complicated by uncertainties over issues such as exposure, toxicity and risk. The vast majority of chemicals entering the environment have not been subjected to rigorous toxicity testing and, in most cases, the only available data are derived from occupational studies where exposures are generally much higher or from animal experiments from which extrapolation to humans is fraught with problems. Neither source reflects probable exposure events associated with soil contamination. In many cases, data from high-dose exposures are extrapolated to the low-dose levels experienced by the public around contaminated land. Such a technique, carried out linearly, inevitably highlights some level of risk even at the lowest exposures

---

[1] Y.I. Baris, L. Simonato, R. Saracci and R. Winkleman, in *Geographical and Environmental Epidemiology*, eds. P. Elliott, J. Cuzick, D. English and R. Stern, Oxford University Press, Oxford, 1996, p. 310.

Issues in Environmental Science and Technology No. 16
Assessment and Reclamation of Contaminated Land
© The Royal Society of Chemistry, 2001

despite the evidence of a threshold effect for many chemicals. The public and pressure group may then, incorrectly, interpret this as meaning that no exposure is safe.

In addition, identifying viable routes of exposure to soil contaminants can be extremely difficult and even when exposure exists it is often of low magnitude, difficult to quantify or insignificant. To further exacerbate such difficulties, there are few environmental standards available for soil and very little work has been carried out on the effects of environmental exposure to mixtures of chemicals. Undoubtedly, the dangers associated with exposure to toxic substances in land are less overt than exposure to air pollution such as vehicle emissions but are becoming increasingly important to the public and regulators, particularly given the current demand for housing and land in the UK. Exposure to soil contamination can and does occur and may occasionally be sufficient to elicit an effect.

2 Exposure

Estimating Exposure

Understanding how people are exposed to soil contaminants is fundamental to any study investigating a possible link with health. That said, most health and epidemiological studies fail to appreciate, or identify, if and how people are exposed. For example, many studies of communities around landfill sites simply use proximity to the site of interest as a proxy measure of exposure. Often such an approach is unavoidable, as there may be insufficient data to allow for a proper exposure assessment. While this approach has some merit, the lack of a proper exposure assessment will mean that the outcomes of the study (whether positive or negative) will be significantly weakened.

Such uncertainties relating to exposure have been acknowledged in the recently introduced contaminated land legislation (see chapter by Lowe and Lowe) which describes the need for a pollutant linkage to be identified before land can be considered contaminated. In any investigation of the potential health effects of contaminated land, it is important that viable pollutant linkages are, at least, identified and, if possible, quantified to some extent. Even where an excess of disease may exist, if a specific hazardous exposure cannot be identified then no link can be made.

Ideally, a health risk assessment would characterise the dose–response relationship, *i.e.* the relationship between the dose of a chemical administered or received and the incidence and/or severity of an adverse health effect in an exposed population. However, estimating the dose–response relationship for many chemicals (particularly environmental agents) is often extremely difficult or, because of the lack of data, unachievable. For example, little is known about dermal uptake rates of soil-bound contaminants or the duration of such contact episodes. Therefore, estimating the dose received from dermal contact with soil can be highly tentative and is usually based upon a number of simplifying assumptions.

Even where dose–response relationships are available, they have severe limitations when used in environmental exposure settings. Most relate to acute exposure to relatively high concentrations of a chemical, whereas in reality, concentrations of a chemical within the soil environment are likely to be much lower. Often the effects of continuous low-level exposure to a chemical in the environment are difficult to define and the situation may be complicated by the presence of other chemicals. To date, the cumulative effects of low-level exposure to a number of environmental agents are poorly understood.

It is also important to recognise that human responses to chemicals in the environment will be extremely varied and some people will be particularly sensitive, for example children and people with pre-existing disease. Therefore, any available or derived dose–response relationship or model is unlikely to take into account the inherent heterogeneity of the exposed population.

This chapter will predominantly deal with the risks from chemical contaminants but the reader should be aware that soils contain radioactive substances (*e.g.* radon) and a variety of biological pathogens (*e.g.* tetanus, anthrax) that can adversely affect human health.

Exposure Pathways

When air and water become polluted, it is relatively easy to identify the most important exposure pathways (*i.e.* for water it would typically be consumption). However, humans may be exposed to soil contaminants through several potential exposure routes and the significance of each of these routes will vary depending on the contaminant involved, the geological characteristics of the soil, the type of receptor, the site location and the proposed end use of the site. It is important to ensure that all potentially viable exposure pathways are properly evaluated. Fundamentally, soil contaminants can enter the body by three main pathways:

- consumption of soil, food or water;
- inhalation of soil particles or vapours;
- dermal contact (including puncture).

Time and space do not allow the detailed review of all potential exposure pathways but Table 1 briefly summarises many of the potential routes of exposure to soil contaminants. However, it is worth briefly considering the potential for exposure through what many believe to be the most important pathway for the general public, the ingestion of soil. It is widely accepted that contaminant intake from soil ingestion at residential sites may exceed intakes from all other pathways combined.[2] Soil may be ingested *via* several routes. It may be ingested directly, either as the result of deliberate ingestion, or from accidental/casual hand to mouth ingestion. Soil can adhere to plant surfaces either through 'soil splash' or harvest practices[3] with the result that particles of soil may be ingested attached to food items and poorly washed food may contain

[2] C. C. Ferguson, *Land. Contam. Reclam.*, 1996, **4**, 171.
[3] S. C. Sheppard and W. G. Evenden, *Environ. Geochem. Health*, 1992, **14**, 121.

Table 1 Examples of potential routes of exposure to soil contaminants

Exposure route	Example
Ingestion of soil	Accidental or deliberate soil ingestion by an infant
Ingestion of dust	Accidental ingestion of household dust containing soil
Ingestion of contaminated food	Consumption of home-grown vegetables
Ingestion of contaminated drinking water	Consumption of contaminated groundwater
Inhalation of soil particles	Inhalation of contaminated dust generated during remediation of contaminated land
Inhalation of soil vapours	Inhalation of volatile chemical from soil
Dermal contact with soil	Contact with soil during remediation of contaminated land
Dermal contact with dust	Contact with dust-containing soil
Dermal contact with contaminated water	Contact with contaminated water through washing

significant quantities of soil. In addition, very fine grains of soil may become embedded in plant tissues and may not be remove by washing, preparation or cooking.[2]

Most concern centres on the potential for exposure in young children, particularly those between the ages of 1 and 5 years. Such children are most vulnerable because their soil intake rates will be substantially higher than in older children and adults and because young children are inherently more sensitive to soil contaminants. There are a number of reasons why young children will consume more soil than adults. Typically children between the ages of 1 and 5 can be expected to accidentally ingest as much as 40–200 mg of soil per day because they spend more time playing outside (hence have a greater contact time with soil), have a poorer concept of hygiene (children have little or no sense of disgust) and will exhibit mouthing behaviours such as finger sucking, hand in mouth activity *etc.* Of most concern is the fact that most young children will at some time or another experiment with deliberate soil ingestion. During early development, all children will demonstrate mouthing behaviours and many will experiment with eating or chewing on non-edible objects including soil. Typically this is a short-lived exploratory behaviour but even short episodes of deliberate soil ingestion can result in quite dramatic soil ingestion rates. However, in some children, this behaviour may not be short-lived and such children can often exhibit a condition know as soil pica or geophagia (a desire to eat soil). Such children are particularly at risk as soil intakes can be extremely high. Recent estimates suggest that geophagic children may ingest anywhere from 1–8 g of soil per day and in extreme cases as much as 25–60 g per day.[4] Such ingestion rates have potentially serious health implications and Figure 1 illustrates the difference between accidental soil ingestion in an adult and typical soil ingestion rates for infants and geophagic children. Obviously geophagia can present significant

4 E. Calabrese, E. J. Stanek, R. C. James and S. M. Robert, *Environ. Health Perspect.*, 1997, **12**, 1354.

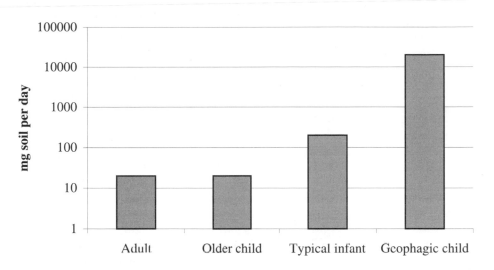

cause for concern. Edward Calabrese and colleagues[4] elegantly illustrated the potential seriousness of soil ingestion in the USA. They showed that supposedly conservative risk based soil standards which are commonly thought of as being protective of health may not be fully protective of children exhibiting acute soil pica (geophagia) behaviour. In addition to such concerns, geophagia is also an acknowledged risk factor to toxocariasis (infestation in humans of the larvae of dog and cat roundworms), particularly in children living in developing countries.

Once a child reaches the age of about 6–7, soil intake rates would be expected to decrease to around 10–100 mg soil per day, estimates that take into account accidental ingestion from hand to mouth activities and ingestion of soil adhering to food. Most risk assessment methodologies employ a soil ingestion rate of 200–400 mg soil per day for children and 5–10 g soil per day for children displaying soil pica, values which are reasonably reflective of probable soil intake rates for the majority of people. However, as recent research suggests, soil intake rates during pica episodes can be higher and it must be recognised that many assumptions of soil intake may not be wholly protective of particularly susceptible population sub-groups. Furthermore, many risk assessments often ignore the risk to geophagic children because the condition is considered rare.

While geophagia is mostly confined to young children below the age of three, adults may also display geophagia and during pregnancy women may crave soil (clay). The significance of high soil consumption by pregnant women has not been elucidated but the potential for harm must be considered high. Geophagia is also commonly reported in less developed countries where the eating of soil may satiate extreme hunger and compensate for mineral deficiencies.[5] There is also good evidence that in certain parts of the world the reason for deliberate soil ingestion by children and adults may be cultural. For example, in many parts of

[5] M. A. Oliver, *Eur. J. Soil Sci.*, 1997, **48**, 573.

the Deep South in the USA the eating of soil (typically clay) is an acquired social behaviour and bags of clay can often found on sale in general stores.

Contaminant Uptake

As previously mentioned, there are great uncertainties associated with estimating exposure and contaminant intake and it is worth briefly discussing the further difficulties in assessing contaminated uptake. Intake rates are not the same as uptake rates (many people often confuse the two) and the potential dose received by an individual at the point of exposure is not, of course, the actual internal dose taken up by the target organs *etc*. Not surprisingly, given our limited understanding of exposure, there are few robust data on uptake rates of soil-bound contaminants. Yet such an understanding is vital in determining dose–response relationships. Currently there is much research into this area, particularly examining uptake rates from soil ingestion and dermal contact. Some of the difficulties associated with such research can be highlighted using ingestion of soil-bound contaminants as an example.

In order to evaluate exactly how much ingested contaminant is taken up in the gut, many researchers have compared uptake rates for soil-bound contaminants with uptake rates of the same contaminant in food. Not surprisingly, current research indicates that uptake rates from soil are not necessarily consistent with contaminant uptake rates from food or with the pure, soluble forms usually employed in animal studies. Data on the relative bioavailability of soil contaminants are limited but some studies suggest that uptake rates for soil contaminants are lower than those for food or water because strong soil–contaminant bonds will reduce the amount of contaminant available for uptake. However, data are available which suggest the opposite and reports of higher uptake rates for some soil contaminants are often ascribed to the fact that contaminants bound to soil are retained for longer in the gut because soil particles become entrained in the gut mucosa. This would, in theory, increase the time for uptake and this hypothesis is supported by well-documented agricultural practices. Farmers have long supplemented nutrient feeds for cattle with soil to maximise nutrient uptake. Obviously such variations are important when assessing the risk from soil ingestion.

Uptake rates may also differ with age. Young children, for example, absorb lead *via* the gastro-intestinal tract four to five times more efficiently than adults, which makes them more susceptible to the effects of lead. Certainly much work remains to be done on soil contaminant uptake rates but current research illustrates that uptake rates are highly variable and dependent on a range of factors such as the contaminant type, physical and chemical properties of the soil, residence time in the gut *etc*.

Targets (Who Is Most at Risk?)

At many sites (*e.g.* residential developments with domestic gardens, open spaces) children are particularly sensitive receptors or targets. This is because children:

- are more susceptible to environmental toxins because their systems are still developing (including growth and development, immature body organs and tissues and, during infancy, a more vulnerable immune system);
- spend a greater proportion of their time outdoors;
- are exposed to different or greater environmental hazards than adults as a result of their behaviour, *e.g.* crawling and play activities increase the potential for exposure to soil;
- have a poorer concept of hygiene and have a natural curiosity and tendency to explore their surroundings;
- consume proportionately more food and drink and breathe more air per kg of body-weight than adults.

However, it is important to acknowledge other particularly sensitive targets. Workers employed during the remediation of contaminated sites are also particularly at risk from soil contamination, and good on-site hygiene and health and safety cannot be emphasised enough, as it is important that all potential exposure routes are minimised or prevented. Indeed, remediation of such land can, if not properly controlled, pose a great challenge to the health of both residents and workers. There are a number of cases when workers have suffered adverse health effects when working on contaminated ground. For example, in the USA, workers excavating a new subway tunnel under a former petrol station were exposed to petrol fumes when they inadvertently dug into soil contaminated by petrol.[6] Reported symptoms included nausea, headache, throat and eye irritation and cough. More recently, workers employed to remediate and develop the site of the Millennium Dome in London, which was (and to some extent still is) heavily contaminated with a range of toxic chemicals, suffered dermatitis after contact with contaminated soil. On-site health and safety must identify all potential hazards and take appropriate steps to reduce or prevent exposure.

Gardeners, particularly allotment owners, are another group of potential receptors who may be at increased risk if contamination is present within the ground. In such cases, exposure *via* dermal contact, accidental soil ingestion or the food chain may be important. The latter is especially true when home-grown produce makes a significant contribution to their diet. This is unlikely for most people in the UK because we receive the majority of our food from a large number of separate and geographically diverse sources. However, allotment owners will often consume much of what they grow and, therefore, home-grown produce may make a significant contribution to the diet and contaminant intake through the diet may be high. Furthermore the emergence of 'city farms' in the last decade, whereby urban communities are encouraged to grow their own produce, while extremely worthy enterprises, may put users at risk if such farms are located on polluted urban soils.

Does Exposure Occur?

Some of the inherent difficulties in estimating exposure have been highlighted

[6] A. L. Davidoff, P. M. Keyl and W. Meggs, *Arch. Environ. Health*, 1998, **53**, 183.

but, because exposure tends not to be obvious and, in most cases, is seldom high enough to cause acute health effects, it is important to look for another indicator of exposure. Analysis of tissues and body fluids for chemicals and their metabolites, enzymes and other biological substances can provide evidence of the interaction of environmental chemicals with biological systems. Biomarkers can be used to assess the exposure and effect of chemicals and also to confirm the exposure of individuals or populations to a particular substance. The presence of biomarkers does not necessarily mean the individual will suffer an adverse health effect and often they identify alterations in tissues and body fluids that do not result in measurable health effects. However, biomarkers can provide persuasive evidence that exposure to soil contaminants can occur and are widely used to identify and quantify exposure.

For example, exposure to lead can be identified through analysis of blood and such analysis has revealed that soil lead can often be an important source of exposure, especially in young children, and can remain a major source of increased lead absorption for many years. Several studies have demonstrated a relationship between proximity to contaminated land and blood lead, with highest blood lead levels in children living on or near contaminated sites. The results of studies around metal smelters indicate that after closure strong relationships between blood lead levels in children and proximity to the smelter still exist, suggesting that contaminated soil rather than air may now be the primary source of exposure. As a result it has been suggested that elevated soil lead can independently predict blood levels in children, and undoubtedly mouthing, pica or geophagia can be significant transfer routes. However, it is likely that soil brought indoors as dust is a more important source of exposure for the majority. Undoubtedly in young children, lead in soil and dusts can be important routes of exposure, but the importance of these will be expected to diminish with age. In most cases, it is unclear whether air lead *per se* or fallout lead in dust or soil is the major contributor to blood lead.

In the USA, the Center for Disease Control (CDC) and Agency for Toxic Substances and Disease Registry (ATSDR) believe that there is evidence that elevated blood lead levels in children may be associated with elevated concentrations of lead in soil and dust,[7] particularly when the concentration exceeds $500-1000 \, mg \, kg^{-1}$. A study in Birmingham, UK has shown dust lead loading in the home to be an important predictor of blood lead concentrations in children[8] while soil lead in older areas of Sydney, Australia is considered to represent a significant health hazard to young children.[9]

Similar relationships have been reported for other heavy metals. For example, urinary levels of arsenic and cadmium have been associated with contaminant levels in soils. However, in all these studies it is important to appreciate that potential confounding from other environmental sources makes confirming soil as an independent predictor of heavy metals in the body difficult. Finally, it is

---

[7] Agency for Toxic Substances and Disease Registry, *The Nature and Extent of Lead Poisoning in Children in the United States*, A Report to Congress, Washington, DC, 1988.

[8] I. Thornton, D. J. Davies. J. M. Watt and M. J. Quinn, *Env. Health Perspect.*, 1990, **89**, 55.

[9] M. J. Fett, M. Mira, J. Smith, G. Alperstein, J. Causer, T. Brokenshire, B. Gulson and S. Cannata, *Med. J. Aust.*, 1992, **157**, 441.

reassuring that despite many urban soils in the UK containing lead concentrations above 500 mg kg^{-1}, blood lead levels in most children and adults are, in fact, very low.

3 Health Studies

Biomarkers can demonstrate that exposure to soil contaminants does occur in many cases, but it is extremely difficult to demonstrate a cause and effect relationship because the interaction between humans and environmental agents may not elicit a noticeable (measurable) health effect. Furthermore, there may be the confounding effects of exposure from other sources and other substances and from socio-economic factors such as deprivation.

The Royal Commission on Environmental Pollution reported in 1996 that it was not aware of any study that provided firm evidence of adverse effects of contaminated land on health.[10] However, the potential for health effects is real. Exposure to chemicals is a risk factor for several diseases and the International Agency for Research on Cancer has classified over 70 chemicals as human carcinogens, 60 as probable and more than 200 as possible carcinogens.[11] There is evidence of an environmental influence in human diseases such as cancer,[12] including geographic variation in cancer incidence in migrant populations, occupational studies and animal experiments. A number of studies also appear to suggest an association between other health effects and exposure to environmental chemicals. There are thousands of sites contaminated with hazardous chemicals with the potential to expose human populations. In the USA, the ATSDR has estimated that 46% of the assessed National Priority List sites represent an actual hazard to health.[13]

However, there is actually little published evidence to draw on and even if controversial evidence is accepted, the total burden of disease that can be potentially attributable to environmental contamination will be modest in comparison with the range of avoidable disease. However, this 'absence of evidence' is different from 'evidence of absence' and such an approach does not consider potential inequalities in risk distribution.

Some of this uncertainty is due to inherent weaknesses of environmental epidemiological studies, which cannot provide an entirely convincing level of scientific confidence. Epidemiological research is relatively inefficient in identifying the relationships between health and the environment. The overwhelming majority of studies in this field have no direct measure of exposure and many fail to appropriately control for potentially powerful confounding factors such as deprivation and other environmental exposures. There may be prolonged latency or exposure periods that need to be considered and population migration is difficult to take into account. In addition, many of these studies have statistical weaknesses such as multiple testing (the more statistical tests that are carried out

[10] Royal Commission on Environmental Pollution Nineteenth Report, *Sustainable Use of Soil*, HMSO, London, 1996.
[11] International Agency for Research on Cancer, url: http://www.iarc.fr/.
[12] E. G. Peeters, *Experientia*, 1987, **43**, 74.
[13] B. L. Johnson and C. DeRosa, *Rev. Environ. Health*, 1997, **12**, 235.

the more likely it is that a spurious association will emerge simply by chance) and low numbers. However, there is a growing literature and some potentially productive research themes are emerging.

The real level of health risk needs to be considered and, while there is clearly potential for exposure to chemicals through human interaction with soil, it is important to put such exposure into perspective. People experience an enormous range of chemical exposures in ordinary day-to-day living and ultimately there is no such thing as a 'no-risk' environment. In addition, it is difficult to quantify the health consequences of the aesthetic and social qualities of the environment (*e.g.* leisure and recreation opportunities) that contamination may affect.

Physical Effects

Contaminated land can present a physical hazard to site users. Landfill gas can cause explosion and fire. Explosions have occurred under houses built near to former landfill sites in Loscoe (Derbyshire) and Kenilworth (Warwickshire). Sites may also present physical dangers such as shafts, holes, tanks, tunnels, dilapidated structures and subsidence due to collapse of underground workings. The colliery spoil slide at Aberfan in 1966 killed 144 people including 116 school children.

Carcinogenic Risk

Concerns that soil is implicated in cancer development are not new and date back to the concept of 'cancer houses' in the 19th century. Studies from mid-19th century England and Wales suggested higher cancer rates in low-lying clay areas and lower rates in limestone or chalk soils. However, most studies show a link between cancer and a deficiency in soil chemistry rather than a chemical excess.[12,14] There is a strong positive relationship between natural radiation (radon) and lung cancer, and pleural cancer has been linked with exposure to natural sources of asbestos-like fibres in Turkey.[1]

There is little consistency in the literature on residential proximity to contaminated sites and cancer incidence, and while some studies fail to demonstrate an effect,[15,16] some investigations of the geographic variation in cancer incidence/mortality have indicated a link with residential proximity to hazardous waste sites.[17] While some of these associations have limited biological (*e.g.* female breast cancer) or temporal plausibility, many landfill sites certainly contain extremely hazardous substances, including known carcinogenic compounds.

Perhaps the most studied landfill site is the Love Canal site in Niagara Falls, New York State. Between 1920 and 1953 part of Love Canal was used extensively as a dumping ground for a range of extremely toxic waste, including organic pesticide residues. Later, the area was sold and formed part of a residential neighbourhood which included a school. Public concerns were raised in 1978

[14] M. G. Kibblewhite, S. J. Van Rensburg, M. C. Laker and E. F. Rose, *Environ. Res.*, 1984, **33**, 370.
[15] M. P. Dunne, P. Burnett, J. Lawton and B. Raphael, *Med. J. Aust.*, 1990, **152**, 592.
[16] D. B. Baker, S. Greenland, J. Mendlein and P. Harmon, *Arch. Environ. Health*, 1988, **43**, 325.
[17] G. R. Najem and T. W. Greer, *Prev. Med.*, 1985, **14**, 620.

when it became clear that a chemical leachate from this site had made its way into the basements of many nearby houses. The resulting site investigations confirmed excessive chemical contamination of soil. Over 200 chemicals including known or strongly suspected carcinogens such as benzene (a known genotoxic carcinogen), lindane, dioxins and a number of halogenated hydrocarbons were detected. This caused considerable public, professional and political concern with the result that hundreds of people were relocated. A detailed study of cancer incidence around this site did not, with the exception of elevated respiratory cancers, identify any evidence of excess cancers in 'exposed' communities.[18] While rates for respiratory cancer were significantly higher in both male and female subjects, a more detailed analysis of the Love Canal data demonstrated that these rates were not statistically different from those reported in other areas of Niagara Falls which were not influenced by this site. In other words, the excess was part of a wider area and was not simply associated with Love Canal.

Cancer incidence and mortality was examined in the area around the Castlereagh Regional Waste Disposal depot, New South Wales, Australia which had received over one million tonnes of liquid waste and an unspecified range of solid domestic industrial and hazardous waste since 1974.[19] This study considered indirectly age/sex standardised ratios in areas within 3 km of the site using cancer data for the period 1974–1998 and mortality data for the period 1975–1992. No excess cancer or mortality rates were found except for brain cancer but this excess could not be confirmed on reanalysis of the data.

Rates of cancers have also been examined around landfill sites in the UK. One study compared observed cancer rates in children living near a landfill site in Walsall[20] with the average regional rate over the period 1980–1988. The community was considered to be potentially exposed to airborne pollutants and from possible underground seepage of toxic wastes. Comparison of observed and expected cancer rates in these areas did not show any statistically significant differences, although the small sample size reduced the statistical power of the study. Additionally, there was no direct measure of exposure and no control of confounding factors such as poverty. Another study (also in the West Midlands) investigated cancer rates in a population living in the vicinity of a closed landfill site in Wolverhampton.[21] Despite considerable public anxiety, no evidence of an alleged cancer cluster was found.

One of the difficulties in these studies is establishing whether the study population has been or is actually being exposed to something hazardous and most simply use the proximity approach discussed earlier. However, occasionally a study has some plausible likelihood of a real exposure to known carcinogens. One such study was that by Goldberg and colleagues[22] who examined cancer

[18] D. T. Janerich, W. S. Burnett, G. Feck, M. Hoff, P. Nasca, A. P. Polednak, P. Greewald and N. J. Vianna, *Science*, 1981, **212**, 1404.

[19] A. Williams and B. Jalaludin, *Aust. New Zeal. J. Pub. Health*, 1998, **22**, 342.

[20] K. R. Muir, J. P. Hill, S. E. Parkes, A. H. Cameron and J. R. Mann, *Paediat. Perinatal Epidemiol.*, 1990, **4**, 484.

[21] P. J. Saunders, A. J. Kibble, R. L. Smith and L. Somervaille, *Report on Cancer Incidence in the Area around Weddell Wynd 1986–1995*, Institute of Public and Environmental Health, University of Birmingham, 1998.

[22] M. S. Goldberg, N. al-Homsi, L. Goulet and H. Riberdy, *Arch. Environ. Health*, 1995, **50**, 416.

rates in communities around an actively gassing municipal solid waste landfill site in Montreal, Quebec. This site contains about 36 million tons of domestic, commercial and industrial waste and was still operating at the time of the study. The composition of the gas produced by this landfill site included a range of volatile organic compounds including recognised carcinogens such as benzene and vinyl chloride and several suspected carcinogens such as methylene chloride, 1,4-dichlorobenzene and carbon tetrachloride. A population based cancer registry was used to identify incident cases. Cancer rates were compared with non-exposed reference areas in Montreal, which were similar to the exposed areas in terms of several key sociodemographic factors. After adjustment for age and calendar year, some significantly elevated risk ratios were identified but there was no evidence of a dose–response effect and these findings could have been as a result of multiple testing (148 statistical tests).

Although there is no consistent evidence of a link between residential proximity to a contaminated site and cancer incidence, it is important to acknowledge that a number of studies have tentatively reported possible associations and this will cause concern to local populations worried about perceived environmental contamination.

Reproductive Effects

In addition to studies examining cancers, many researchers have also sought to examine the possibility that living near landfill sites/contaminated land can have reproductive effects. Studies into the potential effects of contaminated sites on reproductive health tend to be inherently stronger than cancer studies because they avoid the uncertainty associated with long lead times between exposure and the onset of disease. The latency period for many reproductive effects is short (weeks or months) when compared with cancers, which can take years or even decades to appear. While some studies have been inconclusive or shown no reproductive effects,[23–25] the larger and better conducted studies generally have identified some cause for concern.

Congenital Anomalies. Congenital anomalies relate to malformations present at birth and, therefore, exposure around the time of conception or during pregnancy are considered most important. At present, little is known about risk factors for congenital anomalies and causes are divided into three categories: unknown, genetic and environmental. Several studies have examined environmental factors and most have focused on parental occupational exposure. However, published literature in this area is contradictory and inconsistent, although exposure to organic chemicals, including chlorinated solvents and pesticides, radiation and electromagnetic fields are certainly suggestive of an association.

Several researchers have also examined environmental exposure before and

[23] S. G. Shaw, J. Schulman, J. D. Frisch, S. K. Cummins and J. A. Harris, *Arch. Environ. Health*, 1992, **47**, 147.

[24] W. A. Sosniak, W. E. Kaye and T. M. Gomez, *Arch. Environ. Health*, 1994, **49**, 251.

[25] E. G. Marshall, L. J. Gensburg, D. A. Deres, N. S. Geary and M. R. Cayo. *Arch. Environ. Health*, 1997, **52**, 416.

during pregnancy and several studies suggest that residence near hazardous waste sites may be associated with increased risk of congenital anomalies. In the USA, proximity to hazardous waste sites in New York State has been suggested to carry a small additional risk of congenital anomalies.[26] This study examined over 9000 congenital malformations and 17000 healthy controls living in the vicinity of 590 hazardous waste sites. Using a hazard rating system from the US Environmental Protection Agency (USEPA) to clarify exposure risk, and after controlling for several confounding factors, the study found a small but significant additional risk of giving birth to a child with a congenital anomaly in mothers living closer to a hazardous waste site.

Recently, two further studies have reported similar findings.[27,28] The EUROHAZCON study examined the risk of congenital anomalies near 21 hazardous waste sites using data from seven regional registers of congenital anomalies in five European countries (including the UK).[27] All the landfills studied contained hazardous waste of non-domestic origin. Around each study site an area of 7 km radius was defined, which included a zone of 3 km radius thought to represent the area of most exposure (called the 'proximate' zone). However, these zones are essentially arbitrary and in both the 7 km and 3 km zones other forms of environmental exposure would be expected. These potentially confounding factors were not taken into account during the study.

For every case, controls were randomly selected from all children without malformations born on the nearest following day in the same study area. All cases and controls were geographically located with the address or postcode of the mother's place of residence, with an accuracy of 100 m, and residence was used as a proxy for exposure to chemical contaminants from the landfill site.

Residence within 3 km of a landfill site (all study areas combined) was associated with a statistically significant raised risk of congenital anomaly after adjustment for maternal age and socioeconomic status. There was also a fairly consistent decrease in risk of congenital anomalies with increasing distance from the hazardous waste sites. However, when each study area was examined separately this association was not consistently found and many study areas reported no excess risk. This is not surprising, as each landfill site will be different in terms of its composition, topography, engineering and management; factors which will greatly influence the potential for exposure to hazardous materials. In addition, there is no evidence-based rationale for the choice of study areas and no account taken of meteorological influences on the level of exposure.

In South Wales, the Welsh Combined Centres for Public Health compared indices of health in a population living near a landfill site with a population matched for socioeconomic status and examined environmental monitoring data.[28] Using odour complaints as a proxy for exposure in five wards near the site, the authors compared mortality, rates of hospital admission and measures of reproductive health (proportion of all births and stillbirths of infants weighing

[26] S. A. Geschwind. J. A. Stolwijk, M. Bracken, E. Fitzgerald, A. Stark, C. Olsen and J. Melius, *Am. J. Epidemiol.*, 1992, **135**, 1197.

[27] H. Dolk, M. Vrijheid, B. Armstrong, L. Abramsky, F. Bianchi, E. Garne, V. Nelen, E. Robert, J. E. Scott, D. Stone and R. Tenconi, *Lancet*, 1998, **352**, 423.

[28] H. M. Fielder, C. M. Poon-King, S. R. Palmer, N. Moss and G. Coleman, *Br. Med. J.*, 2000, **320**, 19.

< 2500 g; rates of admissions for spontaneous abortion; rates of all reported congenital malformations) within 22 wards in the same local authority with similar deprivation levels. The study found no consistent differences in mortality, rates of hospital admissions or proportion of low birthweight infants between the two populations. However, there was a significantly increased maternal risk of having a baby with a congenital abnormality in residents near the site, both before and after its opening, although a cluster of cases of a congenital anomaly (gastroschisis) post-dated its opening. To date the authors have been unable to confidently link residential proximity to the site with health effects to the local population and have recommended further studies. At the time of writing there are at least two large studies currently examining congenital anomalies around landfill sites around the UK.

Low Birth Weight.   A statistical excess of low birth weights among Love Canal residents between 1940 and 1953 has been reported. Vianna and Polan studied the incidence of low birth weights among white live-born infants from 1940 to 1978 in various areas of Love Canal.[29] Examination of adverse reproductive events such as low birth weight can avoid some of the problems associated with the uncertainty in latency periods between exposure and clinical diagnosis. In the areas adjacent to the site, 6.5% of live births were low birth weight infants, whilst in the area considered to have been most exposed (based on probable drainage pathways) 12.1% had low birth weights. The latter was significantly higher than the comparison area (7.8%) and in other residents of the canal area. Potential confounding factors such as medical-therapeutic histories, smoking, education, maternal age, birth order, gestation period, and urban–rural differences did not appear to account for this observation.

Excess risks for low birth weight and small for gestational age have also been reported in the zone closest to the Montreal landfill site discussed earlier.[30]

Other Health Effects

Another concern associated with Love Canal was the possibility that chemicals within this site may have or had caused chromosomal damage. These concerns were first raised when a study commissioned by the USEPA in 1980 appeared to demonstrate abnormally high chromosomal aberrations in the blood cells of residents that could result in an increased risk of spontaneous abortions and birth defects.[31] Such conclusions alarmed many local residents and in response the US Department of Health and Human Services (HHS) set up a team to review the USEPA's pilot study. The review highlighted several design faults and the study was heavily criticised for its 'alarming conclusions [which were] drawn from data that showed no significant differences from the control data'.[31] The HHS reassessment failed to find any statistical differences between the subjects and their controls for chromosomal damage. However, because the results failed to

[29] N. J. Vianna and A. K. Polan, *Science*, 1984, **226**, 1217.
[30] M. S. Goldberg, L. Goulet, H. Riberdy and Y. Bonvalot, *Environ. Res.*, 1995, **69**, 37.
[31] S. Wolff, *J. Am. Med. Assoc.*, 1984, **251**, 1464.

reassure local residents, the CDC in Atlanta commissioned a carefully controlled study of 46 current or past residents, including 36 people originally tested by the USEPA.[32] This study focused on residents that were considered to have been maximally exposed and each subject was matched with controls from areas in Niagara Falls that did not contain toxic dump sites. The results showed no increase in the frequency of chromosomal aberrations among Love Canal residents but it must be acknowledged that the statistical power of the study is limited because of the difficulty in assessing the residents' past exposure to the chemicals in Love Canal and the small sample size.

Several studies have examined rates of consumption of medication and 'self-reported' symptoms which have been reported to be elevated in populations living near landfill sites. For example, Zmirou and colleagues[33] examined drug consumption in a population in the vicinity of a landfill site after its closure and found no relationship between medication use and estimated exposure, and overall drug consumption did not decrease after the site closed. In another study, Najem and colleagues[34] compared self-reported symptoms in a population living near a hazardous waste site in New Jersey with a non-exposed control population. The study identified exposed populations using intelligence from the enforcement agency and geohydrological information. The study also excluded residents who had lived in the area for less than five years to reduce the effect of migration. Once potential confounders were controlled for, there was no difference in symptom rates.

It is possible that high rates of self-reported symptoms may be due to stress or greater symptom recall. For example, there are reports that elevated symptom reporting rates have continued despite interim remediation or closures of the landfill sites. It is often not clear whether subjective symptoms are the result of a direct toxicological effect, stress, such as environmental or odour worry, or increased symptom perception and recall.

Environmental Worry, Anxiety and Community Concern

The existence of concern itself is known to exacerbate community anxiety. A pollution incident affecting the River Severn in 1994 resulted in tens of thousands of people receiving contaminated drinking water. A subsequent epidemiological study found that reported symptoms were actually related to the ability to detect the contamination (by smell or taste) rather than any exposure itself.[35]

It is often difficult to determine whether symptoms are related to a direct toxicological effect or to stress or worry. For example an investigation of a community living near a hazardous waste site in Queensland, Australia found no evidence of elevated levels of any serious physical disease but high levels of stress and anxiety.[16] Odours may serve as sensory cues for the development of

[32] C. W. Heath Jr, M. R. Nadel, M. M. Zack Jr, A. T. L. Chen, M. A. Bender and J. Preston, *J. Am. Med. Assoc.*, 1984, **251**, 1437.

[33] D. Zmirou, A. Deloraine, P. Saviuc, C. Tillier, A. Boucharlat and N. Maury, *Arch. Environ. Health*, 1994, **49**, 228.

[34] G. R. Najem, T. Strunck and M. Feuerman, *Am. J. Prev. Med.*, 1994, **10**, 151.

[35] S. Fowle, C. Constantine, D. Fone and B. McCloskey, *J. Epidemiol. Commun. Health*, 1996, **50**, 18.

stress-related illnesses or lead to heightened awareness of underlying symptoms. The interaction between symptom prevalence and odour-worry has been examined in a retrospective follow-up of three large epidemiological studies on communities living near hazardous waste sites.[36] On review it was decided that of the reported symptoms, headache and nausea could potentially be related to autonomic or stress-induced mechanisms, while eye soreness or irritation and throat soreness or irritation could be indicative of toxic irritative processes. Prevalence data from 109 respondents were stratified by self-reported frequency of odour perception and by self-reported degree of environmental worry. Respondents were also asked if environmental worry could be ascribed to an illness in themselves or their family (secondary worry).

Significant positive associations were observed between the prevalence of each index symptom (headache, nausea, eye and throat irritation) and both frequency of odour perception and degree of worry. Elimination of 'secondary worry' respondents did not affect these relationships. Relationships were strongest for worry rather than odour. In addition to their individual effects, odour and worry exhibited a positive interaction that, in the case of headaches, was almost multiplicative. The authors concluded that worry and odour perception had 'a potential role' in the development of symptoms near hazardous waste sites.

Another study has examined whether environmental worry, rather than a direct toxicological effect, was the most likely cause of reports of excess symptoms around landfills.[37] This study compared the health of a community after interim clean-up measures of the landfill were introduced with past results on the health of the same community. It concluded that while the exposed population continued to experience and/or report significantly more symptoms than a comparison area, symptom reporting was clearly associated with perceived environmental risk rather than a causal mechanism between exposure and symptoms.

4 Case Studies

While an examination of epidemiological evidence indicates that there is, at present, no consistent or particularly convincing evidence linking contaminated land/landfill sites with health effects, such studies should not be taken as conclusive proof. There are, in fact, a number of cases where contaminated land has had a serious effect on health but given the potentially large numbers of contaminated sites throughout the globe, such cases reflect a very small percentage. That said, they serve as an important reminder of the effects of contaminated land and we shall briefly describe some of these cases.

[36] D. Shustermann, J. Lipscomb, R. Neutra and K. Satin, *Environ. Health Perspect.*, 1991, **94**, 25.
[37] J. A. Lipscomb, L. R. Goldman, K. P. Satin, D. F. Smith, W. A. Vance and R. R. Neutra, *Environ. Health Perspect.*, 1991, **94**, 15.

Respiratory Cancer and Contaminated Soil in the Cappadocian Region of Turkey

It became apparent in the 1970s that people living in the rural Cappadocian region of Turkey had a particularly high incidence of respiratory cancers, including the relatively rare condition pleural mesothelioma; a respiratory disease strongly associated with exposure to asbestos dust. Subsequent environmental investigations demonstrated that the soils of this region contained high levels of erionite fibres for which there is sufficient evidence in both humans and animals of carcinogenicity. Subsequent epidemiological studies have demonstrated a very strong association between residence in this area and respiratory cancer (especially pleural mesothelioma) with the main exposure route being inhalation of contaminated soil.[1]

Cadmium Polluted Soil

Jinzu Valley, Japan. One of the most infamous cases of contaminated land and health occurred in Japan and the effects were most prominent immediately after the Second World War. Around the end of the 19th century, soils in the Jinzu River basin, part of the Toyama prefecture, became contaminated with cadmium as a result of activities upstream at the Kamioka mines. The main activity at this mine was the mining and processing of zinc (cadmium is often associated with zinc ores) with the result that wastewater rich in heavy metals was discharged into the Jinzu River. Contaminants from this industry moved down-stream and caused contamination of soils in paddy fields as a result of abstraction of river water into fields in order to cultivate the local rice crop. Under favourable conditions, cadmium can be a fairly mobile heavy metal, particularly in soils with low pH, and increases in soil cadmium can often result in an increase in the uptake of cadmium by plants. This in turn results in an increase in dietary exposure and the consumption of contaminated agricultural crops can be a major pathway of human exposure.

This is what happened in the Jinzu River basin as locally grown rice became contaminated with cadmium. Dietary exposure was high as rice was the staple diet of most local people and as a result of chronic exposure to cadmium in the diet many began to suffer bone disease and abnormalities of calcium metabolism and kidney problems, a condition that known locally as itai-itai disease (literally translated as 'ouch ouch').[38] This disease became most apparent during and immediately after the Second World War with elderly women, particularly women who had given birth to several children, most affected. During this period, food was scarce and the diets of local people became extremely heavily dependent on rice. In addition, the scarcity of food meant that the diet of many people was also deficient in calcium, vitamin D and protein. These factors exacerbated the toxicity of cadmium, and in a relatively small area over two hundred elderly women were severely affected, many of whom eventually died as a result of this condition. It has since been estimated that people in the Jinzu River basin had a

---

[38] *Cadmium—Environmental Aspects*, Environmental Health Criteria Monographs No. 135, World Health Organization, Geneva, 1992.

cadmium intake more than 10 times the maximum tolerable daily intake.

Other rural populations in Japan have been similarly affected, particularly communities living in the cadmium polluted areas of the Ishikawa and Akita Prefectures. In all cases, those individuals with itai-itai disease or who had severe renal dysfunction as a result of environmental exposure to cadmium have a distinctly poor life expectancy.

Shipham.   A similar situation arose in the UK when a national survey of stream sediments in 1978 revealed extremely high levels of cadmium, lead and zinc in Shipham, Somerset,[39] where zinc and lead mining had been carried out in the 19th century. Analysis of locally grown vegetables found that cadmium levels were also elevated and typically ranged from 5 to 20 times above normal levels,[40] while in certain leafy vegetables, such as spinach, cadmium levels were up to 60 times normal. In fact, recorded concentrations of cadmium in soil in the Shipham area were far higher than those recorded in the Jinzu valley and there was great concern that such contamination could have an effect on the health of local people. As a result, a programme of studies was conducted to assess the body burden and any associated health effects.[41]

Biomarker studies provided evidence of increased cadmium exposure. There were small but measurable increases in cadmium levels in urine and blood samples in Shipham compared with a control population but in most cases such measurements fell between accepted 'normal' ranges. Despite such increases, there was no evidence that exposure had been sufficient to elicit a measurable effect. For example, there was no significant excess of symptoms in Shipham compared with a control area and no significant differences in blood pressure (cadmium has been linked to hypertension).[41] In addition an analysis of hospital admissions and cancer registrations revealed that any effect of cadmium was considered unlikely.

Some small excesses of some cadmium related disorders (genitourinary disease, nephritis, hypertension and cardiovascular disease) were identified, but the study concluded that if cadmium did have an effect on mortality it was slight. Recent findings from a cohort study of Shipham that covered 60 years (up to 1997) provides further reassurance.[42] This study found that all cause cohort mortality in both Shipham and the comparison population in the village of Hutton were lower than expected (using the South West Region as a whole). Although there were apparent excess mortalities for a number of diseases including cerebrovascular disease and hypotension in the Shipham area, these were not statistically significant and could therefore simply be explained by chance alone. This study did find some tentative evidence of excess prostate cancers in the Shipham cohort

[39] H. Morgan, in *The Shipham Report. An Investigation into Cadmium Contamination and Its Implications for Human Health*, ed. H. Morgan, *Sci. Total Environ.*, 1988, **75**, 11.

[40] B. J. Alloway, I. Thornton, G. A. Smart, J. C. Sherlock and M. J. Quinn, in *The Shipham Report. An Investigation into Cadmium Contamination and Its Implications for Human Health*, ed. H. Morgan, *Sci. Total Env.*, 1988, **75**, 41.

[41] C. Strehlow and D. Braltrop, in *The Shipham Report. An Investigation into Cadmium Contamination and Its Implications for Human Health*, ed. H. Morgan, *Sci. Total Environ.*, 1988, **75**, 101.

[42] P. Elliott, R. Arnold, S. Cockings, N. Eaton, L. Järup, J. Jones, M. Quinn, M. Rosato, I. Thornton, M. Toledano, E. Tristan and J. Wakefield, *Occup. Environ. Health*, 2000, **57**, 94.

but the numbers involved were too low to have any statistical power.

The lack of real evidence of adverse health effects in Shipham has been ascribed to the fact that locally grown vegetables would form only a low percentage of the diet, since most food consumed by the UK population is derived from areas remote from the point of consumption. In addition, when compared with the population in the Jinzu Valley, the diet of the local people in Shipham was varied and generally good.

5 Conclusion

Public concern about the effects of contaminated land is growing and regulatory authorities, researchers and health officials need to address such concerns. While it seems obvious to determine exactly what people are being exposed to, many studies are undertaken where there is little or no evidence of any viable route of exposure. In such cases it may be extremely difficult to determine whether any perceived or actual health effect is due to soil contamination or is simply the result of confounders such as environmental worry, or recall bias. Currently there is very little evidence that contaminated land in the UK has caused ill-health but global evidence demonstrates that, when exposure occurs, contaminated land can and does have a serious impact. However, the one common factor in most of these cases is that the actual level of exposure is relatively high (*e.g.* locally grown rice in Jinzu Valley was the main food source; in Shipham it was not) whereas on most contaminated sites the actual level of exposure will be expected to be much lower. It is also worth appreciating that a lack of evidence does not mean that there is no effect. The absence of an established link between contaminated land and health in the UK may also be, in part, the result of limitations in risk assessment and epidemiological techniques.

Recent studies linking residence near hazardous waste sites with increased risk of congenital malformation are ample evidence that the potential threat from contaminated land must not be ignored. Although these studies are, in parts, inherently weak and do not provide conclusive evidence of a link with contaminated land, they certainly warrant further, more detailed research. However, it is important that in the future health studies take into account the public health needs of the population under investigation. As we have discussed, concern may trigger ill-health and a well thought out policy of informing the public about the outcomes of any investigation may help relieve fears. This should involve the concerned community in the planning and execution of any study and ensuring that people are kept regularly informed of developments. The ATSDR and UK Department of Health have produced excellent guidance on this issue.[43,44]

Finally it is worth acknowledging the increasing awareness, at both public and regulatory levels, of contaminated land issues in the UK and other developed countries. In 1995 the West Midlands Surveillance System for chemical incidents

[43] Agency for Toxic Substances and Disease Registry, *A Primer on Health Risk Communication Principles and Practices*, Agency for Toxic Substances and Disease Registry, Georgia, 2000.

[44] Department of Health, *Communicating about Risks to Public Health: Pointers to Good Practice*, The Stationery Office, London, 1998.

was established to monitor and record environmental contamination events that caused or had the potential to cause public health problems. In the first two years, contaminated land issues barely rated a mention and were dwarfed by air and water issues. However, the subsequent three years have seen a dramatic increase in contaminated land related incidents, an increase that reflects both increased awareness in contaminated land and also concern from the general public about such sites. No doubt such concern will increase in the coming years and it is the task of epidemiologists, regulators and public health officials to ensure that such concerns are adequately and properly investigated.

Human Health Risk Assessment: Guideline Values and Magic Numbers

C. PAUL NATHANAIL AND NAOMI EARL

1 Land Contamination Policy

The UK has adopted a risk-based approach to managing historic land contamination. The *Guidelines for Environmental Risk Assessment and Management*, colloquially referred to as Green Leaves II,[1] is a revision of earlier guidance[2] that had become a standard reference, encouraging the public and private sectors to take environmental risks into account during decision making. The revised guidance emphasises the establishment of risk assessment, risk management and risk communication as essential elements of structured decision making across Government. Much more emphasis has been placed on the importance of public participation in environment risk management. The guidance provides an overarching framework and serves as a 'first port of call' for many regulators before they tackle specific risk assessment processes. The following stages of risk assessment are promoted:

- Problem formulation
- Risk assessment, incorporating:
  hazard identification
  identification of consequences
  assessment of the magnitude and probability of those consequences
  assessment of the significance of the risks ensuing
- Options appraisal
- Risk management

UK Government policy with respect to land contamination was laid out in *Framework for Contaminated Land*.[3] The Framework has informed Government

---

[1] Department of the Environment, Transport and the Regions, *Green Leaves II, Guidelines for Environmental Risk Assessment and Management*, prepared by the Institute for Environmental Health, the Stationery Office, London, 2000.

[2] Department of the Environment, *Green Leaves, A Guide to Risk Assessment and Risk Management for Environmental Protection*, HMSO, London, 1995.

[3] Department of the Environment, *Planning Policy Guidance Note 23, Planning and Pollution Control*, London, 1994.

Issues in Environmental Science and Technology No. 16
Assessment and Reclamation of Contaminated Land

action and resulted in the wide adoption of a risk-based approach to contaminated land management, *i.e.* the mere presence of historic pollutants (the hazard) is not sufficient reason to warrant remedial action; some harm to a specified receptor must be possible.

The UK follows the widely recognised contaminant–pathway–receptor concept for assessing risks from contaminated land.[4-8] This contaminant–pathway–receptor relationship is called a pollutant linkage.[7] Receptors may, for example, be humans, surface water, groundwater, buildings, protected ecosystems or property, including livestock and crops.

Ferguson[9] has reviewed recent research in human health risk assessment. For there to be a risk to human health, there has to be a source of contamination, one or more pathways along which the contaminants can travel and one or more humans who may be affected by that contamination. If any element of the linkage (contaminant or pathway or receptor) is missing then the contamination is not regarded as posing a risk.

Human health risk assessment involves identifying and characterising contamination hazards on sites; defining contaminant mobility; developing realistic scenarios in which humans could be exposed to contaminants; estimating the uptake of a contaminant by humans and finally evaluating the consequences of the uptake.

In the continued absence of the DETR/Environment Agency Model Procedures (in preparation) current guidance on land contamination related to (re)development includes:

- PPG 23[3]
- ICRCL 59/83 2nd Edition[10]
- National House-Building Council Chapter 4.1[6]
- EA/NHBC Guidance[8]

PPG 23

Planning Policy Guidance Note 23: Planning and Pollution Control[3] sets out the Government's policies on planning and pollution control. It states that the planning system should not be operated so as to duplicate controls that are the

---

[4] M. Harris, S. Herbert and M. S. Smith, *Remedial Treatment for Contaminated Land, Vol. III: Site Investigation and Assessment*, Special Publication No. 103, CIRIA, London, 1995.

[5] S. Herbert, *United Kingdom*, in *Risk Assessment for Contaminated Sites in Europe, Vol. 2: Policy Frameworks*, eds. C. C. Ferguson and H. Kasamas, LQM Press, Nottingham, UK, 1999, ISBN 0-95330-901-0.

[6] NHBC, *NHBC Standards*, Chapter 4.1, *Land Quality—Managing Ground Conditions*, NHBC, Milton Keynes, UK, 1999.

[7] Department of the Environment, Transport and the Regions, *Circular 2/2000, Contaminated Land*, 2000.

[8] Environment Agency/NHBC, *Guidance for the Safe Development of Housing on Land Affected by Contamination*, Environment Agency R & D Publication 66, the Stationery Office, London, 2000, ISBN 0-11310-177-5.

[9] C. C. Ferguson, Assessing Human Health Risks from Exposure to Contaminated Land, *Land Contam. Reclam.*, 1996, **4**, 159–170.

[10] ICRCL, *Guidance on the Assessment and Redevelopment of Contaminated Land*, ICRCL Guidance Note 59/83, 2nd Edn., Department of the Environment, London, 1987.

statutory responsibility of other bodies (including the Environment Agency). PPG 23 sets out current guidance to Local Authorities on how to deal with land contamination through planning. Planning controls are not an appropriate means of regulating the detailed characteristics of potentially polluting activities.

DETR is preparing further planning guidance on land contamination which will amplify the guidance in PPG23, explain the interface with the regime under Part IIA of the Environmental Protection Act 1990 and provide local planning authorities with technical and practical advice on land contamination issues. In the meantime, advice in PPG 23 remains valid (see ref. 7, para. 47).

ICRCL 59/83 2nd Edition

The Interdepartmental Committee on the Redevelopment of Contaminated Land (ICRCL) Guidance Note 59/83 2nd Edition[10] presents a systematic procedure for the assessment of contaminated land. The Guidance Note is intended for use in assessing land that is to be redeveloped and not to assess sites already in use. Conversely the contaminated land regime contained in Part IIA of the Environmental Protection Act 1990 is concerned with the current, not intended, use of sites. Two sets of subjectively determined but nevertheless risk-based trigger concentrations were put forward by the ICRCL to assist in the assessment of the need for remedial action or further investigative work: the lower concentration was termed the threshold level and the higher the action level. Contaminants present at concentrations below the 'threshold' concentration 'can usually be disregarded, because there is no significant risk that the hazard(s) will occur'. For contaminants present at concentrations above the 'action' level 'the risks of the hazard(s) occurring are sufficiently high that the presence of the contaminant has to be regarded as undesirable or unacceptable. Action of some kind, ranging from minor remedial treatment to changing the proposed use of the site entirely, is then unavoidable.' For contaminants present at concentrations above threshold but below action levels 'there is a need to consider whether the presence of the contaminant justifies taking remedial action for the proposed use of the site. If such considerations suggest that some action is justified, then it should be taken; the decision to do so is therefore based on informed judgement.'

Substances were classified into Group A: Contaminants which may pose hazards to health, Group B: Contaminants which are phytotoxic but not normally hazardous to health and a third group of other inorganic and organic substances and soil conditions. Action levels were never published for Group A and B contaminants, resulting in many regulators and contaminated land consultants wrongly interpreting exceedance of the threshold values as meaning that remediation is required. As shown above, this was never the intention of the document's authors.

NHBC Chapter 4.1 Land Quality—Managing Ground Conditions

The NHBC offers a ten year warranty for newly built homes to NHBC registered builders and developers and purchasers of their houses—the *Buildmark* warranty.

Buildmark covers the full cost, if it is more than £500 at April 1999 prices, of cleaning up contamination of the land on which the house is built. This new cover will operate if a local authority or the Environment Agency takes action under Environmental law. NHBC Chapter 4.1[6] was issued in order to enable the *Buildmark* warranty to be extended to cover statutory remediation of contaminated land. Chapter 4.1 provides a framework for managing geotechnical and contamination issues to ensure:

- all sites are properly investigated and assessed using a source–pathway–receptor methodology (*sic*);
- foundations and substructure designs are suitable for the ground conditions;
- sites are properly remediated where necessary or appropriate design precautions are taken, and appropriate documentation and validation can be provided to NHBC.

Environment Agency/NHBC

The Environment Agency[8] guidance for the safe development of land affected by contamination provides advice on: identifying sites where contamination may exist; collating data; consulting with regulatory and planning authorities; designing and executing ground investigations; applying generic assessment criteria to contaminants in soil; site specific risk estimation; selecting and implementing a remedial strategy; validating and monitoring remediation; and managing information and reporting.

The report gathers together best practice in relation to building housing on contaminated land for the first time ever. It provides practical guidance, in an easy to follow, step-by-step format and contains flow charts including a fold-out for reference.

2 Risk Assessment

Exposure assessment involves modelling, as appropriate and relevant, contaminant uptake *via* ingestion, inhalation and skin contact routes. At its simplest, risk assessment, therefore, consists of comparing total daily uptake (*i.e.* from both soil and background sources) with a tolerable daily uptake as advised by authoritative toxicologists.

The approach taken for carcinogens and non-carcinogens differs. Carcinogens are assessed on the basis of excess lifetime cancer risk arising from the soil of the site being considered. The effects of more than one carcinogen should be treated as being additive (SEPA 2000). Non-carcinogens are assessed on the basis of a chronic safe daily intake that allows for background exposure to the contaminant. The effects of more than one non-carcinogen should be treated as additive if they affect the same end point.[11]

---

[11] SEPA, *Methodology to Derive Tolerable Daily Intakes.* Contract No. NO230/2940, 1999. Available at www.sepa.org.uk/contaminated.land/links/tdi.pdf

Procedures for the systematic estimation of human health risks from exposure to contaminated soil have developed rapidly over the past 10–15 years. Reduced to its simplest form the overall procedure of a conceptual model driven site specific risk assessment can be reduced to eight steps:

1. Determine the representative concentration of contaminant in soil.
2. Determine or assume site exposure characteristics taking into account current and intended land use.
3. Conduct exposure assessment (through gut, lung and skin routes as appropriate) for defined receptors.
4. Derive intake of contaminant from site soil (cautiously assume intake is equivalent to uptake).
5. Determine intake of contaminant from other sources (for non-cancer end points only).
6. Calculate total intake from site soil and other sources from 4 and 5 above.
7. Determine tolerable daily intake for contaminant.
8. Compare total intake with tolerable daily intake.

The exposure characteristics and assumptions refer partly to site conditions (*e.g.* soil type, soil chemistry, exposed soil fraction) and partly to the site use and users (*e.g.* construction style of buildings, receptor characteristics, activity patterns and time spent on site). Many of these are assumptions rather than measurements because chronic health risk assessments are concerned with predicting contaminant intakes and their effects up to 70 years into the future. Some of the soil characteristics may also change with time.

Exposure assessment involves modelling contaminant intakes *via* ingestion, inhalation and skin contact routes. Some of the algorithms currently used are complex, and some may be difficult to validate. Allowance may also need to be made for intake of the same contaminants from background, especially dietary, sources. In practice, because data on uptake (absorbed dose) are rarely available, the comparison is often between estimated intake and tolerable daily intake (TDI). This is precautionary because soil contaminants are not usually as bioavailable as the pure and soluble compounds typically used in the animal studies conducted to derive tolerable daily intakes.

3   What Is a Guideline Value?

Although Part IIA recommends the use of guideline values by a Local Authority, the term is not defined in the Statutory Guidance.[7] Guideline values provide nationally consistent guidance on the likely need for soil remediation without the very substantial costs associated with developing site specific assessment criteria.[12] In the UK the CLEA model, described below, is being used to derive such guideline values in a rolling programme that will over the next few years eventually cover some fifty or so substances.

---

[12] C. C. Ferguson and J. Denner, *Soil Guideline Values in the UK*, in *Contaminated Soil '93*, eds. Arendt *et al.*, pp. 365–372.

Guideline values have embedded in them policy decisions on acceptable levels of risk which may have been made taking costs and benefits into account.[13]

Why Are Guideline Values Useful?

It should be noted that contaminants behave differently and occur in different forms depending on the medium in which they occur. Guideline values will vary for any given substance depending on whether contamination of soil, water or gas is being assessed. A panel of European experts working within the EU funded Concerted Action for Contaminated Sites (CARACAS) reported that 'A combined approach using guideline values to streamline the preliminary stages of decision making and site specific risk assessment to achieve fine-tuning in later stages of an investigation, is generally considered the most appropriate'.[14] Guideline values, if appropriately used, can reduce the cost of risk assessment and simplify decision making. They are easy to understand and interpret by a wide variety of non-specialist stakeholders.

Why Are Guideline Values Dangerous?

Unlike a fish out of water, a table of guideline values separated from its explanatory text does not die. Instead it takes on a new, unintentional and often undesirable life of its own. The numbers therein take on a significance far in excess of that for which they were intended: contexts are lost, boundaries are forgotten and interpretations of exceedance become irrelevant. In the attempt to speed up, simplify or reduce the cost of risk assessment the 'magic number' reigns.

For example in recent years in the UK exceedance of the ICRCL threshold levels for metals and arsenic have been wrongly interpreted as indicating the need for remediation.

What Do Other Countries Do?

Many other countries have produced their guideline, trigger, action, intervention or screening values for a variety of purposes. It is currently common practice in the UK to use the Dutch values[15] for substances where ICRCL[10] provides no trigger values. On occasions other national or regional 'magic numbers' are used. This section provides a brief introduction to the background of commonly encountered European and North American values.

It cannot be overemphasised that the use of non-UK values should only be countenanced if on a site-specific basis the values can be shown to be appropriate and protective of the relevant receptor(s) of concern.

[13] C. C. Ferguson and J. Denner, Developing Guideline (Trigger) Values for Contaminants in Soil: Underlying Risk Analysis and Risk Management Concepts, *Land Contam. Reclam.*, 1994, **2**, 117–123.

[14] C. C. Ferguson, D. Darmendrail, K. Freier, B. K. Jensen, J. Jensen, H. Kasamas, A. Urzelai and J. Vegter, eds., *Risk Assessment for Contaminated Sites in Europe, Vol. 1: Scientific Basis*, LQM Press, Nottingham, UK, 1998.

[15] VROM, *Circular on Target Values and Intervention Values for Soil Remediation*, Ministry of Housing, Spatial Planning and Environment (VROM), the Netherlands, DBO, 2000, ISBN 1-99922-686-3.

Italy. The Italian National Agency for Protection of the Environment (ANPA) and the Regional Agencies for the Environment are undertaking an intensive programme of developing technical approaches to risk assessment and restoration of contaminated sites.[16] Within this programme a risk-based tiered procedure for decision-making has been laid out. This procedure envisages two simplified types of risk-based assessment for site screening and defining clean-up objectives: generic and site-specific criteria. 'Generic acceptability limits' or guideline values are defined according to a land-use sensitive risk assessment approach. Different receptors and exposure pathways are considered through validated exposure and contaminant migration models. Groundwater is protected as a drinking water resource by using regulatory water standards and appropriate soil values. Site-specific risk assessment is encouraged to ensure that clean-up objectives and remedial actions are appropriate for local exposure conditions. Guideline values for residential/recreational and industrial/commercial land uses have been developed for 94 substances in soil and 92 in groundwater.[16-18]

Norway. To assist in the regulation of contaminated sites in Norway a two-tiered decision model was developed and put into force in 1995.[18] This model focused on the need to address the sources, pathways and effects of contamination on human health, groundwater, nearby surface water (including fjords), and the soil environment. Generic target values were developed for a number of metals, metalloids and organic substances or groups of substances. These target values, which relate only to the most vulnerable land use, are based on the then current Danish and Dutch values for contaminated sites. For other land uses, or for occasions when target values are exceeded, a system of site-specific risk assessment was applied. New guideline values were published in 1999.[16,18,19] The values relate to a child exposed to contamination through soil and dust ingestion, dermal contact, dust and vapour inhalation, drinking on-site groundwater and consumption of vegetables grown on the site.

Sweden. The Swedish EPA has developed guideline values for 36 contaminants or contaminant groups in soil.[16,20] For each substance guideline values are developed for three different types of land use:

1. Land with sensitive use, *e.g.* residential areas, childrens' nurseries, agriculture, together with groundwater abstraction
2. Land with less sensitive use, *e.g.* offices, industries, roads, car parks, but still with groundwater abstraction
3. Land with less sensitive use as above but with no groundwater abstraction

[16] URL CLARINET, 2001, www.clarinet.at/policy.

[17] *Technical Regulation for Remediation of Contaminated Sites*, DM n471/99, Italian legislation, 1999.

[18] C. C. Ferguson and H. Kasamas, eds., *Risk Assessment for Contaminated Sites in Europe, Vol. 2: Policy Frameworks*, LQM Press, Nottingham, UK, 1999, ISBN 0-95330-901-0.

[19] E. Vik, G. Breedveld and T. Farestveit, *Guidelines for the Assessment of Contaminated Sites*, Norwegian Pollution Control Authority, Oslo, 1999, www.sft.no.

[20] Swedish Environmental Protection Agency, *Development of Generic Guideline Values: Model and Data Used for Generic Guideline Values for Contaminated Soils in Sweden*, Naturvardsverket Forlag, Stockholm, www.environ.se.

The generic values have been derived using a Swedish exposure model based on similar models and data developed by other countries and international organisations. Data were chosen, and in some cases adapted, so that the resulting values are appropriate for Swedish conditions regarding geology, exposure, sensitivity and policy. The guideline values indicate concentrations above which 'undesirable effects' on human health and/or the environment may occur. The main report describing how to use the guideline values is written in Swedish, although a summary report describing the model and data used for developing guideline values is published in English.[20] This report warns that the guideline values will not be appropriate for all sites and site-specific assessment criteria will need to be developed in such cases.

Spain.   Spanish regions have developed separate approaches to the management of land contamination—reflecting their disparate industrial history.[16,18]

The Basque Region has published soil screening values, known as Indicative Values for Assessment (VIE), which are land-use dependent and provide a generic assessment criterion that will allow essentially risk-free soils to be differentiated from soils that pose or could potentially pose risks for the intended use. They are applied in the exploratory phase of an investigation and are used by the competent authority to help determine the need for further detailed investigation or, in some cases, immediate remedial action. Three levels have been established. The first, VIE-A, is derived on the basis of concentrations found in natural soils with little anthropogenic input, interpreted as involving no significant risk for any likely soil use. The second, VIE-B, represents the level above which more detailed consideration of risks is required; and the third, VIE-C, is the value above which the risk becomes unacceptable. While the text focuses on the recovery of previously contaminated soil, prevention of future contamination is also addressed.

Netherlands.   The Dutch target and intervention values[15] are commonly used in the UK as generic assessment criteria. They were developed using a Dutch exposure assessment model incorporating assumptions appropriate to Dutch site conditions, including soil organic matter content and clay content, and lifestyles. VROM[15] includes correction factors for soil organic matter and clay content that, unfortunately, are rarely applied in the UK. Target values for approximately one hundred substances were generated in 1987 to support the implementation of the Soil Protection Act (Wbb). Soil contaminant concentrations below the target value were deemed to pose no risk and the site was deemed to be multi-functional. Target values were checked against practical feasibility and define a relatively unpolluted soil/sediment. In the context of remediation, achieving the target value implies the recovery of the soil's functional properties to support human, plant and animal life.

Under the Wbb contamination caused after 1 January 1987 should be 'cleaned up as quickly as possible, irrespective of the concentrations encountered and the risk of the pollutants'.[15] Historic (pre-1987) contamination still requires management and the Dutch Intervention Values were introduced in 1994 to assist. Intervention values are a numeric value above which there is serious contamination and the functional properties of the soil, with respect to humans, flora and fauna, have been compromised or are in danger of being so. Target and intervention values apply to both soils and sediments. Where no intervention value has been set a value indicative of serious contamination is provided. Since there is considerable uncertainty associated with the indicative levels, they are not accorded the same status as intervention values. Nevertheless soil concentrations in excess of the indicative values require a regulatory response. For those substances for which no number exists, such as asbestos, current guidance provides a number of alternative approaches (Annex D of VROM[15]).

The 'Dutch numbers' have changed several times since they were first published. The latest version is available from the VROM web site.

USA. There are a number of federal and state sets of guideline values. The US EPA federal soil screening levels were developed to help standardise and accelerate the evaluation and clean-up of contaminated soils at sites on the National Priorities List (NPL).[21] The US EPA are clear that the soil screening levels (SSLs) are not national clean-up targets and exceeding them does not necessarily trigger the need for remedial action. Rather they were developed to screen out areas of a site on the NPL that do not require further action or study under the Comprehensive Environmental Response, Compensation and Liability Act (CERCLA)—better known as the Superfund. The SSLs anticipate a future residential land use.

The toxicological basis of SSLs are oral cancer slope factors, non-cancer reference doses, inhalation unit risk factors and reference concentrations. Drinking water and health based standards are used to determine screening levels in groundwater.[21]

4 The Use of Guideline Values

Part IIA of the Environmental Protection Act 1990 requires Local Authorities to identify contaminated land. Circular 2/2000[7] provides Local Authorities with advice on how to determine that 'there is a significant possibility of significant harm being caused'. The following extracts refer to specific indicated sections:

'B.45 The local authority should determine that land is contaminated land on the basis that there is a significant possibility of significant harm being caused (as defined in Chapter A), where:

(a) it has carried out a scientific and technical assessment of the risks arising from the pollutant linkage, according to relevant, appropriate, authoritative and scientifically based guidance on such risk assessments;
(b) that assessment shows that there is a significant possibility of significant harm being caused; and
(c) there are no suitable and sufficient risk management arrangements in place to prevent such harm.

[21] United States Environmental Protection Agency, *Soil Screening Guidance: Technical Background Document*, OSWER 9355.4-17A, EPA/540/R-95/128.

B.46 In following any such guidance on risk assessment, the local authority should be satisfied that it is relevant to the circumstances of the pollutant linkage and land in question, and that any appropriate allowances have been made for particular circumstances.

B.47 To simplify such an assessment of risks, the local authority may use **authoritative and scientifically based guideline values** for concentrations of the potential pollutants in, on or under the land in pollutant linkages of the type concerned. If it does so, the local authority should be satisfied that:

(a) an adequate scientific and technical assessment of the information on the potential pollutant, using the appropriate, authoritative and scientifically based guideline values, shows that there is a significant possibility of significant harm; and

(b) there are no suitable and sufficient risk management arrangements in place to prevent such harm.

B.48 **In using any guideline values, the local authority should be satisfied that:**

(a) the guideline values are relevant to the judgement of whether the effects of the pollutant linkage in question constitute a significant possibility of significant harm;

(b) the assumptions underlying the derivation of any numerical values in the guideline values (for example, assumptions regarding soil conditions, the behaviour of potential pollutants, the existence of pathways, the land-use patterns, and the availability of receptors) are relevant to the circumstances of the pollutant linkage in question;

(c) any other conditions relevant to the use of the guideline values have been observed (for example, the number of samples taken or the methods of preparation and analysis of those samples); and

(d) appropriate adjustments have been made to allow for the differences between the circumstances of the land in question and any assumptions or other factors relating to the guideline values.

B.49 The local authority should be prepared to reconsider any determination based on such use of guideline values if it is demonstrated to the authority's satisfaction that under some other more appropriate method of assessing the risks the local authority would not have determined that the land appeared to be contaminated land.'

When Not to Use Guideline Values

Guideline values should not be used where the criteria in B47 and B48 above have not been demonstrably shown to have been met. Under Part IIA the Local Authority will be responsible for assessing the applicability of any given guideline value. For redevelopment purposes that responsibility rests with the developer. Commonly encountered misuse of guideline values includes:

1. Use of ICRCL threshold values as implying the need for remediation.

2. Use of Dutch values without correction for soil composition.
3. Use of Dutch intervention value where receptor driving the value is not human health.
4. Use of total PAH or TPH measurements without reference to the nature of the hydrocarbon being considered.
5. Ignoring the influence of soil conditions such as pH, soil organic matter or particle size distribution.
6. Ignoring the additive effects of exposure to carcinogens.
7. Ignoring the additive effect of non-carcinogens affecting the same end point.

What Does it Mean if You Exceed a Guideline Value?

The speeding analogy: guideline values may be thought of as being analogous to a speed limit—a limit imposed to ensure safety on a given type of road. If a speed limit is exceeded the regulator takes one of the following courses of action:

- Nothing
- Warning
- Licence endorsed and/or fine
- Loss of licence
- Jail

The response depends on the amount of the exceedance and on the regulator's prosecution policy.

If a soil guideline value is exceeded the response could be:

- Nothing—the exceedance is trivial
- Reconsider exposure assessment assumptions
- Remedial action is needed

The response will be influenced by the amount of exceedance, the consequences of exceedance and the nature of the guideline value. Regulators may also apply a prosecution policy which reflects their priorities in the face of limited resources.

How Much Above a Guideline Value Is too Much?

Spatial, sampling and analytical variability will generate considerable uncertainty in any chemical analysis. Guideline values are sometimes developed within given sampling and/or analytical protocols (*e.g.* US EPA[21]). Exposure assessment models are based on many dated, inadequate or inappropriate studies and therefore tend to take a cautious approach. Therefore in general a small exceedance is likely to mean that the critical receptor is facing a risk that is deemed by the appropriate government to be acceptable. However, non-specialists may find this a difficult concept to accept.

For some contaminants exceedance by as much as one or two orders of magnitude can be within the range of probable over-conservatism in the guideline value and the derivation of site-specific assessment criteria should be

95

considered if remediation is to be avoided. Exceedance above this is likely to require remedial action. For smaller exceedance, less than one order of magnitude, it is highly likely that it may be tackled by more site specific measurements, reducing the conservatism in guideline values.

If Soil Concentrations Do not Exceed the Guideline Value Is That OK?

Generic guideline values are designed to be cautious and protective of a very wide range of site conditions and receptor characteristics and behaviour. Therefore in most cases where soil concentrations do not exceed an appropriate guideline value no further action is warranted. Indeed if all the criteria described above for the selection of guideline values have been met then non-exceedance should be interpreted as indicating no need for further action. In practice guideline values may be used inappropriately and non-exceedance may not be protective of the critical receptor. Regulators should ensure that any guideline values used for generic risk assessment have been demonstrated to be applicable on a site specific basis. DETR[7] warns that

'In some cases, the information obtained from an inspection may lead the Local Authority to the conclusion that, whilst the land does not appear to be Contaminated Land on the basis of that information assessed on the balance of probabilities, it is still possible that the land is Contaminated Land. This might occur, for example, where the mean concentration of a contaminant in soil samples lies just below an appropriate guideline value for that contaminant. In cases of this kind, the Local Authority will need to consider whether to carry out further inspections or pursue other lines of enquiry to enable it either to discount the possibility that the land is Contaminated Land, or to conclude that the land does appear to be Contaminated Land. In the absence of any such further inspection or enquiry, the local authority will need to proceed to make its determination on the basis that it cannot be satisfied, on the balance of probabilities, that the land falls within the statutory definition of Contaminated Land.'

5   The CLEA Model

Ferguson and Denner[12] described CLEA as a generic computer model of human exposure to soil contaminants which:

- Uses well established algorithms
- Uses a Monte Carlo routine to provide explicit procedures for handling uncertainty
- Is designed to be easily updated
- Will complement and facilitate the development of site-specific assessment criteria

CLEA is a probabilistic model of human exposure to contaminants in soil. It is being used to derive new UK Guideline Values to support risk assessments carried out for Part IIA of the Environmental Protection Act 1990 and the planning regime. Ferguson and Denner[22] describe the guideline values as 'intervention' or 'action' values. While the guideline values will represent, as far as possible, the state-of-the-art in scientific understanding, developments in toxicology, exposure assessment, and fate and transport mechanisms will necessitate periodic revisiting and revision of the values.

The toxicological and modelling basis of the CLEA guideline values is described in two documents: *CLR9: Contaminants in Soils: Collation of Toxicological Data and Intake Values for Humans* (DETR in preparation); and *CLR10: The CLEA Model Technical Basis and Algorithms* (DETR in preparation). CLR9 describes how the toxicological input to the CLEA model is derived. This includes the way in which, for non-carcinogens, the tolerable daily intake (TDI) and an estimate of background exposure are selected as well as how carcinogenic risks are calculated. CLR10 summarises technical and scientific basis of the CLEA model and includes details of pathway specific algorithms, receptor characteristics (*e.g.* body weight, height, inhalation rates, soil ingestion, vegetable consumption rates, proportion of vegetables that are home-grown), policy decisions and probability density functions used in Monte Carlo sampling.

For each contaminant for which a guideline value is published there are two further documents: a CLR 9 TOX report and a CLR 10 GV report. CLR 9 TOX describes the toxicology of the substance, reviews toxicological understanding of the substance and recommends values of TDI, mean daily intake and carcinogenic risk. A CLR 10 GV report describing the derivation of the guideline values and laying out the values for specific site conditions and land use scenarios.

Land Use Scenarios

CLEA considers a number of land use scenarios. Ferguson and Denner[12] reported that early versions considered:

1. Residential with gardens
2. Residential without gardens
3. Parks, playing fields and open spaces
4. Allotments
5. Commercial and industrial

The final version of CLEA is likely not to consider all of the above. The land use scenario determines which exposure pathways are relevant, who the critical receptor is, and what their activity patterns are. Guideline values are derived for each land use scenario. The Residential with gardens land use scenario is the most sensitive scenario and the Commercial and industrial scenario the least sensitive because of which pathways are considered. As the guideline values are generic, a reasonable worst case is assumed. They may not therefore be suitable for every site. For instance, the Parks, playing fields and open spaces land use has the same

[22] C.C. Ferguson and J. Denner, *UK Action (or Intervention) Values for Contaminants in Soil for Protection of Human Health*, in *Contaminated Soil '95*, eds. van den Brink *et al.*, 1995, pp. 1199–1200.

exposure assumptions as the Residential without gardens scenario (with the exception of an indoor vapour pathway). This is because it is assumed that children in an urban area may use local parkland in the same way they would a back garden. This assumption may not be true of all parks, playing fields and open spaces.

Assumptions in CLEA

Contaminant behaviour varies with soil type and properties, source zone size and depth below ground level and meteorological conditions. For guideline value derivation purposes, a generic site was developed for which the assumptions about the following were made:

1. Soil type
2. Depth below ground level of contamination
2. Width of source zone parallel to the wind direction
4. Wind speed
5. For organic substances, there is no free product

For certain substances and conditions, different guideline values have been produced. For example, guideline values for certain metals and metalloids vary according to pH (which affects the plant uptake pathway), and for organics vary according to the soil organic matter (SOM) content (which affects the plant uptake and inhalation of soil vapours pathways).

The exposure pathways considered in CLEA are:

1. Direct ingestion of soil and soil-derived dust
2. Consumption of home-grown vegetables grown in site soil
3. Ingestion of soil attached to home-grown vegetables
4. Outdoor dermal exposure to soil
5. Indoor dermal exposure to soil-derived dust
6. Outdoor inhalation of soil-derived dust
7. Indoor inhalation of soil-derived dust
8. Outdoor inhalation of soil vapour
9. Indoor inhalation of soil vapour.

The land use scenario and contaminant determine which pathways are considered. The contribution individual pathways make to the overall exposure varies according to the contaminant and the site conditions since contaminant fate and transport depends on physicochemical properties. Martin and Ferguson[23,24] describe the basis for uptake through the consumption of home-grown vegetables.

[23] I. D. Martin and C. C. Ferguson, *Predicting Trace Metal Uptake by Vegetables Grown in Contaminated Soils*, in *Contaminated Soil '95*, eds. van den Brink *et al.*, 1995, pp. 629–630.

[24] I. D. Martin and C. C. Ferguson, *An Approach for Predicting Trace Metal Uptake by Vegetables Grown in Contaminated Soils*, in *Proceedings of the 3rd International Conference on Biogeochemistry of Trace Elements*, CD ROM 105.pdf, INRA, Paris, 1997, ISBN 2-7380-0775-9.

Ferguson and Marsh[25] describe the basis for contaminant uptake through soil ingestion.

The critical receptor depends on the land use scenario. The exposure and averaging periods vary according to who the critical receptor is, and whether the contaminant is being treated as a carcinogen or a non-carcinogen.

For non-carcinogens within residential scenarios the child in its first six years of life is treated as the critical receptor. Lead and cyanide guideline values consider a child in its second year of life for reasons explained in the SNIFFER Framework (Land Quality Management[26]). Children have a high soil ingestion rate and a low body weight to absorb the contaminant intake compared with adults. The Parks, playing fields and open spaces scenario is considered to be similar to a residential scenario. For Allotments exposure has been assumed to be possible in both childhood and adulthood so exposure and averaging time are both for 70 years. For the commercial and industrial scenario no childhood exposure is assumed and exposure is assumed to be possible for the whole of one's working life.

Carcinogens are assessed by estimating the excess lifetime cancer risk and therefore always have an averaging period of 70 years (the biblical life span of three score years and 10). Exposure periods, however, represent reasonable maximum estimates of exposure to the soil at a site.

The CLEA Guideline Values for non-carcinogens consider exposure from other sources of contaminants. Background exposure from non-soil sources, as defined in the mean daily intake (MDI), includes exposure through drinking water, food consumption and inhalation of ambient air. The MDI is subtracted from the TDI. Exposure to contaminated soil from other sources, such as that which may be tracked back to the house, is taken into account. This is done by allocating such exposure a percentage of (TDI − MDI), usually 50%, to the soil allocation factor (SAF).

For carcinogens, background and other exposure is deemed to be accounted for in the national cancer incidence and therefore neither MDI nor SAF is taken into account.

Modelling Variability and Uncertainty

For a number of parameters used within a risk assessment, there is no one single appropriate value to use. Background data quote a range of possible values. This may be due to natural variation within populations (*e.g.* bodyweight, height, amount of contaminant taken up by plants), differences in behaviour (*e.g.* soil ingestion rate, proportion of vegetables which are home-grown) or due to missing data and/or differences and errors in experimental technique.

Deterministic risk assessment tools deal with this uncertainty by using a single point value for each variable. The use of reasonable worst case estimates can lead

---

[25] C. C. Ferguson and J. Marsh, Assessing the Human Health Risks from Ingestion of Contaminated Soil, *Land Contam. Reclam.*, 1993, **1**, 177–185.

[26] Land Quality Management Ltd., *Framework for Deriving Numeric Targets to Minimise the Adverse Human Health Effects of Long-term Exposure to Contaminants in Soil*, SNIFFER, 2000, available from Foundation for Water Research, www.sniffer.org.uk.

to creeping conservatism. For instance, if the 90th percentile of a probability density function is used as the input to a deterministic risk assessment tool for three separate parameters the combination will lead to an assessment that lies at the 99.9th percentile of likelihood (probability of all three parameters being greater than the 90th percentile is given by $1/10 \times 1/10 \times 1/10 = 1/1000$).

An alternative method a risk assessment model may use is to input the full range of possible values for each uncertain parameter, together with the relevant statistical distributions (probability density functions, or PDFs). Such models are known as probabilistic. CLEA is a probabilistic model. The model samples a point value for each parameter each time it is run and calculates an Average Daily Intake for each run. The way in which the distributions are sampled in CLEA is known as Monte Carlo modelling. A large number of simulations (5000) are run in order to make sure that the statistical distributions are accurately represented. Finally the Average Daily Intakes from all the runs are represented as a distribution. A decision is then made about the percentile of the distribution that should be selected for comparison with the Tolerable Daily Intake. This decision is a matter of policy, not of science. In the case of the CLEA guideline values the 95th percentile has been selected. This does not mean that 5% of the UK population on a site would not be protected. Rather the 5% is the tail end of the combination of unlikely probabilities (a highly underweight child eats the maximum amount of soil, and only home-grown produce, all of which takes up the maximum possible amount of contamination).

Special Case I—Cyanide

A relatively small dose of cyanide from a single exposure can be lethal. Cyanide from lower doses from long term exposure is readily metabolised and eliminated from the body. The CLEA guideline values therefore for both free cyanide and complex cyanides are based on short term risk rather than long term risk.

Special Case II—Lead

The critical receptor for lead uptake is the child in its second year of life. It is the amount of lead that is taken up in blood, rather than the total soil concentration that is important. CLEA guideline values are based on predictions of contaminant concentration in soil that will result in a given amount of lead in the blood. The exposure to lead from other sources far exceeds the exposure to soil; the guideline value for lead is based on the idea of proportional reduction. This means that if *e.g.* 10% of the overall exposure to lead comes from soils, only 10% of the reduction required to bring blood lead within acceptable concentrations should be due to the soil.

CLEA guideline values:

1. Do relate to direct human health risks
2. DO NOT relate to groundwater protection
3. DO NOT relate to ecosystem protection
4. DO NOT apply to occupational risks to site workers

5. DO NOT consider acute risks

6   The SNIFFER Framework

In the absence of a CLEA guideline value, the SNIFFER Framework (Land Quality Management 2000a, b) offers an UK exposure assessment model which is applicable in many situations and for many circumstances.

7   Conclusions

Guideline values have an important role to play as generic criteria in cost effective site specific risk assessment. The appropriateness of any set of guideline values should be established prior to their use on any site. This involves having a thorough understanding of their relevance, assumptions and conditions attached to their use. If soil concentrations exceed an appropriate guideline value by a substantial amount remedial action is necessary. On the other hand, deriving site-specific assessment criteria may show such action is not necessary where soil concentrations are only slightly above guideline values.

The probabilistic CLEA model is being used to develop UK guideline values for human health risk assessment. In the interim the SNIFFER Framework (Land Quality Management[26,27]) offers a simple method for determining site specific assessment criteria.

8   Acknowledgements

One individual's contribution to the development and use of guideline values for the assessment of risks to human health from land contamination within the UK, in Europe and further afield has to be acknowledged at the outset. Colin Ferguson was the intellectual driving force behind the development of both CLEA and the SNIFFER Framework. Tragically he died in 1999, before he could see either tool published.

The work on human health risk assessment begun by Colin Ferguson at the Centre of Research into the Built Environment (CRBE) at the Nottingham Trent University continues today through Land Quality Management at the University of Nottingham, where Colin and his team transferred in 1998. The author has the privilege and duty of leading that team today.

The research was funded in the first instance by the Department of the Environment (now DETR) and later by the Environment Agency following the transfer of the DoE R&D Programme to the then newly formed Agency in 1996. Latterly funding from SNIFFER (www.sniffer.org.uk) and SEPA (www.sepa.org.uk) has supported the development of the SNIFFER Framework.

The work at CRBE and LQM on CLEA and the SNIFFER Framework has been carried out by a team including: Colin Ferguson, Jacqui Marsh, Ammar Abbachi, Ian Martin, Naomi Earl, Paul Nathanail, Caroline McCaffrey and

[27] Land Quality Management Ltd., *Sensitivity Analysis of Framework for Deriving Numeric Targets to Minimise the Adverse Human Health Effects of Long-term Exposure to Contaminants in Soil*, SEPA, Stirling, www.sepa.org.uk.

others. Judith Lowe (née Denner) has been a major intellectual driving force and source of encouragement throughout the process.

This paper has been prepared with the help of Land Quality Management staff Amy Barnes and Caroline McCaffrey to whom we are most grateful.

Any errors or omissions are ours and ours alone.

Ecological Risk Assessment under the New Contaminated Land Regime

MICHAEL QUINT

1 Background and Legal Context

After many years of R&D, the UK government has begun to launch details of the approach it requires for the assessment of land under the new contaminated land regime. This regime went live in April 2000, and it requires local authorities, in conjunction with the Environment Agency, to initiate the inspection and, if necessary, remediation of potentially polluted land.

The essence of the regime is that land should be 'fit for purpose' and that substances in, on or under land should not threaten human health or environmental resources such as groundwater and ecosystems. As a result, chemical pollutants on a site are only to be regarded as an issue if they are capable of causing harm to human health or the environment, or if they have already done so.

The actual wording of the legislation states that land needs to be remediated if it:

'appears to the local authority in whose area it is situated to be in such a condition, by reason of substances in, on or under the land, that

- significant harm is being caused, or there is a significant possibility of such harm being caused
- pollution of controlled waters is being, or is likely to be, caused'

where harm is defined as:

'harm to the health of living organisms or other interference with the ecological systems of which they form part and, in the case of man, includes harm to his property'.

As it is so central to any decision on clean-up, the process of assessing the existence of, and potential for, adverse effects from site-based contaminants is perhaps the most important aspect of the new regime. Depending on the results of this process, contaminant levels may be deemed acceptable for a site's current or

Issues in Environmental Science and Technology No. 16
Assessment and Reclamation of Contaminated Land
© The Royal Society of Chemistry, 2001

planned use, in which case no further action is required, or the site may be judged to require remediation, potentially costing millions of pounds.

The way in which sites are to be assessed under the regime is described in government guidance, much of which has already been issued in either draft or final form. Some of this guidance is statutory (*i.e.* mandatory), while other aspects of it are of a more general, non-statutory nature. The statutory guidance can be found at http://www.environment.detr.gov.uk/contaminated/land/index.htm, while the non-statutory elements are available from the DETR's Land Quality Section and the Environment Agency's National Groundwater and Contaminated Land Centre.

2 Harm to Ecological Receptors

The non-human living organisms and ecological systems that are deemed to be relevant for assessment, and therefore worthy of protection, under the new regime are defined in the statutory guidance as follows:

'Any ecological system, or living organism forming part of such a system, within a location which is:

- an area notified as an area of special scientific interest under section 28 of the Wildlife and Countryside Act 1981;
- any land declared a national nature reserve under section 35 of that Act;
- any area designated as a marine nature reserve under section 36 of that Act;
- an area of special protection for birds, established under section 3 of that Act;
- any European Site within the meaning of regulation 10 of the Conservation (Natural Habitats *etc.*) Regulations 1994 (*i.e.* Special Areas of Conservation and Special Protection Areas);
- any candidate Special Areas of Conservation or potential Special Protection Areas given equivalent protection;
- any habitat or site afforded policy protection under paragraph 13 of Planning Policy Guidance Note 9 (PPG9) on nature conservation (*i.e.* candidate Special Areas of Conservation, potential Special Protection Areas and listed Ramsar sites); or
- any nature reserve established under section 21 of the National Parks and Access to the Countryside Act 1949.'

While the above is a long and detailed list, it basically means that any area which has any form of government protection qualifies as an ecological receptor, as do the animals and plants within it. It should be noted, however, that individual members of protected species (*e.g.* great crested newts) are not identifiable as receptors in their own right, when they occur outside of protected areas. Further, unprotected areas, even those with high ecological value, are also not protected from ecological damage due to land-based contamination under the new regime.

The statutory guidance states that *significant harm* to one of the protected areas identified above consists of:

'Harm which results in an irreversible adverse change, or in some other substantial adverse change, in the functioning of the ecological system within any substantial part of that location; or harm which affects any species of special interest within that location and which endangers the long-term maintenance of the population of that species at that location.

In addition, in the case of a protected location which is a European Site (or a candidate Special Area of Conservation or a potential Special Protection Area), harm which is incompatible with the favourable conservation status of natural habitats at that location or species typically found there.'

Taken together and put simply, this means that, in order for land to be designated as contaminated land on grounds of ecological impact, substantial adverse effects have to occur to either a protected area or to the long-term viability of a population of an important species within it, or there has to be a significant possibility that such impacts will occur at some point in the future. The process of assessing land in the context of these requirements is described below.

3 General Approach to the Assessment of Land under the New Regime

The statutory guidance describes how, in general terms, the assessment of land under the new regime should be based on the principles of risk assessment. These principles have been applied in areas such as chemical safety testing for many years and they include a recognition that risk is a function of hazard and exposure. In simple terms, this means that the mere presence of a hazardous substance on a site is not indicative of a need for clean-up. Instead, it is a trigger for further examination of the likelihood of harm occurring, or that such harm has already occurred, prior to a decision on whether or not clean-up is warranted.

Connected with the adoption of a risk-based methodology is the 'source–pathway–target' or 'contaminant–pathway–receptor' philosophy. This philosophy states that, in order for harm to have occurred, or for there to be a risk of harm occurring, contaminants must have been able to reach relevant receptors *via* specific environmental migration pathways. Where such a relationship exists between a contaminant, a pathway and a receptor at a site, a 'pollutant linkage' is said to exist.

In the initial stages of assessing a site, the contaminant–pathway–receptor philosophy can be used to exclude the site from further consideration on qualitative grounds when one or more of the three components is missing. For example, if the contamination is encapsulated in a sealed on-site cell which prevents the ingress or egress of precipitation, vapours and leachate, and animal or plant contact cannot occur, then no pathways exist and hence no pollutant linkages exist. The land cannot, therefore, be classified as contaminated land.

Where a pollutant linkage is identified, an assessment of its significance must take place in order that the site can be designated as contaminated. Depending on the results of such an assessment, a 'significant pollutant linkage' may be identified, leading to a determination that the land is contaminated. This is a critical point, since it means that the simple existence of a pollutant linkage is insufficient grounds for classifying a site as contaminated. Instead, the linkage has to be shown to be significant before such a determination can be made.

In terms of assessing the existence or likelihood of ecological harm from pollutant linkages, the statutory guidance states that:

'In determining what constitutes such harm, the local authority should have regard to the advice of English Nature and to the requirements of the Conservation (Natural Habitats *etc.*) Regulations 1994.'

At the time of writing, English Nature has yet to produce any definitive guidance in connection with the regime. As a result, the review that follows is based mainly on best practice in ecological risk assessment and work performed recently by the Environment Agency.

4 Assessing Whether Significant Harm to Ecological Receptors *Is* Occurring

According to the statutory guidance, an assessor can only consider that significant harm to ecological receptors *is* occurring where he/she:

'(a) has carried out an appropriate scientific and technical assessment of all the relevant and available evidence; and

(b) on the basis of that assessment, is satisfied on the balance of probabilities that significant harm is being caused'

No other information is provided as to what should be done under (a), above, although some form of ecological fieldwork will be required. As for (b), this seems to require that such a study would need to indicate that there is at least a 51% probability that harm is attributable to site contaminants, although again no further details are provided.

There are several techniques that may be useful in determining whether significant harm is occurring, including:

- Observing gross ecological damage, such as animal carcasses and absent or distressed vegetation. The former are generally obvious, while the latter may require the assistance of specialist botanists. It should be noted that, while such harm may be readily observed, it may still be insufficient to meet the test of significance required.
- Observing reduced fecundity *via* assessments of the population structure of individual species.
- Measuring changes in ecological structure and functioning, using diversity or species abundance indices. This requires that a study is carried out to identify, and count, organisms within the affected area and at a suitable control site. Data from such a study are then used in conjunction with specific algorithms to generate numerical scores, or indices, that can be compared with each other in specific ways. The results of the comparison exercise are then used to conclude whether or not the ecology of the two sites

is significantly different. A review of the various indices that are available for this purpose has been conducted by Magurran.[1]

- Inferring that changes have occurred by comparing observed ecological conditions with suitable benchmarks obtained by, for example, computer simulation modelling. This approach has been used in relation to assessing the impact of industrial discharges to freshwater, *via* the River Invertebrate Prediction and Classification System (RIVPACS) approach (Wright *et al.*[2]).

- Observing pollution-tolerant species that have replaced those typically associated with a particular ecosystem. Examples include the polychaete *Capitella capitata*, and lead-tolerant grasses. The former is able to inhabit petroleum-rich estuarine environments that are toxic to similar species, giving it a competitive advantage that allows it to dominate sediment macroinvertebrate infauna. The latter can be of lower nutritional value to primary consumers and, due to its different vegetative characteristics, can adversely affect ground level microclimates.

- Biomonitoring, in the form of measurements of chemical concentrations in living tissue or observations of animal behaviour. In terms of the former, this can be useful since it is suggestive of the bioavailability of on-site contaminants, as well as their abundance, and it is especially appropriate for plants, filter-feeders and predators such as birds of prey. In terms of the latter, work has been conducted on the locomotive behaviour of certain invertebrates and how this can be adversely affected by contamination (Ferguson *et al.*[3]).

- Identifying biochemical markers. These are typically sublethal biological responses to chemical pollution which are not immediately apparent. For example, genotoxic carcinogens can cause measurable DNA adduct formation, while neurotoxins, such as organophosphate pesticides, can cause reductions of cholinesterase levels in invertebrates.

Since substantial ecological impacts have to occur for a site to be designated as contaminated under the new regime, some of the above are likely to be more useful than others. In this respect, observations of ecological damage backed up with quantitative expressions of impacted ecological structure are likely to be most effective. The more subtle techniques, such as biomonitoring, have a role, however, since they may prove to be useful tools in demonstrating cause and effect, and hence the existence of complete pollutant linkages. In addition, they are useful predictors of ecological effects to come, as described below.

[1] A. Magurran, *Ecological Diversity and Its Measurement*, Princeton University Press, Princeton, New Jersey, USA, 1988.

[2] J.F. Wright, D.W. Sutcliffe and M.T. Furse, *Assessing the Biological Quality of Fresh Waters, RIVPACS and Other Techniques*, Freshwater Biological Association, Ambleside, 2000.

[3] C. Ferguson, D. Darmendrail, K. Freier, B.K. Jensen, J. Jensen, H. Kasamas, A. Urzelai and J. Vegter, (eds.), *Risk Assessment for Contaminated Sites in Europe, Volume 1, Scientific Basis*, LQM Press, Nottingham, 1998.

5 Assessing the *Possibility* of Significant Harm Occurring

According to the statutory guidance, a *significant possibility of significant harm* to ecological receptors occurs if:

- 'significant harm of that description is more likely than not to result from the pollutant linkage in question; or
- there is a reasonable possibility of significant harm of that description being caused, and if that harm were to occur, it would result in such a degree of damage to features of special interest at the location in question that they would be beyond any practicable possibility of restoration'

The guidance then states that:

'Any assessement made for these purposes should take into account relevant information for that type of pollutant linkage, particularly in relation to the ecotoxicological effects of the pollutant.'

Further details specify that a local authority should consider that there is a significant risk of significant harm if:

- (a) 'it has carried out a scientific and technical assessment of the risks arising from the pollutant linkage, according to relevant, appropriate, authoritative and scientifically based guidance on such assessments;
- (b) that assessment shows that there is a significant possibility of significant harm being caused; and
- (c) there are no suitable and sufficient risk management arrangements in place to prevent such harm'

In recognition of the fact that assessments of this nature can be extremely difficult and complex, the guidance then states that pre-determined guideline values, presumably in the form of chemical concentrations in soil, may be used in some circumstances, as follows:

'To simplify such an assessment of risks, the local authority may use authoritative and scientifically based guideline values for concentrations of the potential pollutants in, on or under the land in pollutant linkages of the type concerned. If it does so, the local authority should be satisfied that:

- (a) an adequate scientific and technical assessment of the information on the potential pollutant, using the appropriate, authoritative and scientifically based guideline values, shows that there is a significant possibility of significant harm; and
- (b) there are no suitable and sufficient risk management arrangements in place to prevent such harm.'

In using any guideline values, the local authority should be satisfied that:

(a) the guideline values are relevant to the judgement of whether the effects of the pollutant linkage in question constitute a significant possibility of significant harm;

(b) the assumptions underlying the derivation of any numerical values in the guideline values (for example, assumptions regarding soil conditions, the behaviour of potential pollutants, the existence of pathways, the land-use patterns, and the availability of receptors) are relevant to the circumstances of the pollutant linkage in question;

(c) any other conditions relevant to the use of the guideline values have been observed (for example, the number of samples taken or the methods of preparation and analysis of those samples); and

(d) appropriate adjustments have been made to allow for the differences between the circumstances of the land in question and any assumptions or other factors relating to the guideline values.'

Finally, there is a requirement that the use of such values must be withdrawn when they are inappropriate, as follows:

'The local authority should be prepared to reconsider any determination based on such use of guideline values if it is demonstrated to the authority's satisfaction that under some other more appropriate method of assessing the risks the local authority would not have determined that the land appeared to be contaminated land.'

In essence, therefore, the question of whether or not there is a significant possibility of significant harm must be answered using the most appropriate and scientific risk assessment methods available, although the regulator may use simple guideline values as a short-cut, where this is warranted. However, the use of such criteria as a basis for requiring remediation at a specific site must be withdrawn if a more appropriate assessment shows that there is not, in fact, such a risk.

As a step towards the establishment of technical guidance for such ecological risk assessment, the Environment Agency has recently published a draft document entitled *Assessing Risks to Ecosystems from Land Contamination*.[4] This document is the outcome of a study designed to examine and review a number of international decision-making and risk assessment methodologies and to outline a simple ecological risk assessment framework.

6 The Environment Agency's Suggested Approach

The Environment Agency study reviewed ecological risk assessment frameworks from the USA, Australia, Canada and the Netherlands. In addition, it provided examples of guideline values (GVs) that are used in these countries. The outcome of the study is a suggested ecological risk assessment framework, based on a

---

[4] Environment Agency, *Assessing Risks to Ecosystems from Land Contamination*. Draft Report, National Centre for Ecotoxicology and Hazardous Substances, R&D Technical Report P338, Bristol, 2000.

three-tier system, and incorporating the use of GVs. This framework is described in more detail below.

The first tier of assessment recommended in the Agency's research involves the characterisation of the site and the elucidation of a conceptual model of potential contaminants, pathways and ecological receptors. Soil samples are then taken and analysed for the presence of relevant contaminants, following which numerical hazard indices (HIs) or PEC/PNEC ratios are derived (PECs are 'Predicted Environmental Concentrations' and PNECs are 'Predicted No Effect Concentrations'). PEC/PNEC ratios are obtained by simply dividing the measured chemical concentrations by suitable GVs, while HIs are the sums of such ratios for a number of chemicals where the mode of their toxic action is similar. In both cases, numbers greater than unity are suggested as being indicative of a need to proceed to Tier 2, or undertake risk management, while numbers of less than unity indicate that no further action is required. In the event that the PEC/PNEC ratio or HI is one, or close to one, then monitoring is suggested.

The second tier of assessment involves further site characterisation and the direct toxicity testing of soil samples. In terms of the latter, the following *ex situ* tests are recommended for consideration, depending on the site-specific assessment needs:

- acute earthworm test with *E. foetida* or *E. andrei*
- acute arthropod test with *F. candida* or *F. fimetaria*
- root growth test with a monocotyledon such as barley
- dicotyledon root growth test with, for example, cress
- microbial test with *Vibrio fischeri*
- microbial nitrogen mineralisation test
- microbial carbon mineralisation test

In situ tests are also suggested for Tier 2, including:

- bait lamina tests for organic decomposition
- minicontainer tests for organic decomposition
- neutral red retention in earthworm coelomocyte lysosomes

There is also an option to perform biomonitoring, in the form of testing tissue samples for contaminants. As described previously, this can be used to confirm the existence of suspected pollutant linkages.

No details are given on which tests should be performed during Tier 2 and there is no framework for their application at specific sites. Methods for interpreting the results are described in general terms, however, particularly in relation to their extrapolation to population, community and ecosystem effects, although again the details are only sketchy. Clearly this is an important consideration in the light of the requirements of the guidance. It is also suggested that site-specific PEC/PNEC ratios are derived, and that probabilistic uncertainty analyses are performed. Again, methods for doing so are not described in any detail, however.

In the event that Tier 2 indicates that the ecological risk is significant, the assessment either proceeds to Tier 3, or risk management is undertaken. In the event that the risk is shown to be insignificant, no further action is required. As at Tier 1, monitoring is suggested for a borderline situation.

Tier 3 provides the opportunity for further, more detailed analysis of site-specific ecological risk. This may involve the use of similar tools to those used in Tier 2, or it may require the application of more comprehensive methods. Such methods include individual-based population models, the biological testing of a wider variety of species than at Tier 2 and chronic, sub-chronic or mesocosm toxicity tests. If, following such an assessment, ecological risks appear to be sufficient for the site to be designated as contaminated land, then remediation is required. Conversely, the demonstration of an absence of such risk will result in no further action being necessary. As before, a borderline situation will be met with a requirement for monitoring.

7 Guideline Values

As indicated above, the statutory guidance states that guideline values (GVs) can be used to help assess the possibility of significant harm to ecological receptors from site-based contaminants, provided such values meet certain requirements. The Environment Agency study[4] described above identified a number of candidate values from a diversity of sources. These are reproduced in Table 1. This study also refers to a Scottish Environment Protection Agency report that provides details on how to derive soil screening values for the protection of ecological receptors from soil-based contaminants. Further details on the contents of this report are not provided, however.

Table 1 indicates that candidate GVs for a single chemical can vary by up to three orders of magnitude, depending on the origins of the values and their stated protection goals. For example, tin has a microbial benchmark from the US Oak Ridge National Laboratory of 2000 mg kg^{-1}, while the Canadian remediation criterion for this chemical for the protection of agricultural soils is 2 mg kg^{-1}. The variability between such candidate GVs is partly a function of the approaches used in their derivation and partly due to their underlying ecotoxicological data. Regardless of its origin, however, such variability is clearly relevant if GVs from elsewhere are to be used in the UK.

As indicated in the guidance, GVs must only be used where they are appropriate to the site. Examples of situations where they may not be appropriate include:

- chemical forms and states that are different from those used in the ecotoxicological studies
- species and habitats that differ from those on which the GVs are based
- different geological, climate and other environmental settings
- exposure scenarios which can be modelled more accurately by adopting different techniques
- instances where ecotoxicity criteria have been changed or superseded
- situations where different approaches to modelling toxicity are more scientifically valid

Table 1 Soil ecotoxicological benchmark values taken from various sources (taken from ref. 4)

Chemical	ORNL Earthworm Benchmark (mg kg⁻¹)	ORNL Microbial Benchmark (mg kg⁻¹)	ORNL Phytotoxic Benchmark (mg kg⁻¹)	CCME RC (mg kg⁻¹)	RIVM EIVs (mg kg⁻¹)	ANZECC Phytotoxic EILs (mg kg⁻¹)
Aluminium	—	600	50	—	—	—
Arsenic	60	100	10	20	40	20
Barium	—	3000	500	750	625	300
Boron	—	20	0.5	2	—	—
Cadmium	20	20	4	3	12	3
Chromium	0.4	10	1	750	230	400
Cobalt	—	1000	20	40	240	—
Copper	60	100	100	150	190	100
Fluorine	—	30	200	200	—	—
Iron	—	200	—	—	—	—
Lanthanum	—	50	—	—	—	—
Lead	500	900	50	375	290	600
Lithium	—	10	2	—	—	—
Manganese	—	100	500	—	—	500
Mercury	0.1	30	0.3	0.8	10	1
Molybdenum	—	200	2	5	480	—
Nickel	200	90	30	150	210	60
Selenium	70	100	1	2	—	—
Silver	—	50	2	20	—	—
Tin	—	2000	50	2	—	—
Titanium	—	100	—	—	—	—
Vanadium	—	20	2	200	—	50
Zinc	100	100	50	600	720	200
Phenol	30	100	70	0.01	40	—
4-Nitrophenol	7	—	—	0.01	—	—
3-Chlorophenol	10	—	7	0.05	10	—
3,4-Dichlorophenol	20	—	20	0.05	10	—
2,4,5-Trichlorophenol	9	—	4	0.05	10	—
2,4,6-Trichlorophenol	10	—	—	0.05	10	—
2,3,4,5-Tetrachlorophenol	20	—	—	0.05	10	—
Pentachlorophenol	6	400	3	0.05	5	—
Chlorobenzene	40	—	—	0.1	30	—
1,4-Dichlorobenzene	20	—	—	0.1	30	—
1,2,3-Trichlorobenzene	20	—	—	0.05	30	—
1,2,4-Trichlorobenzene	20	—	—	0.05	30	—
1,2,3,4-Tetrachloro-benzene	10	—	—	0.05	30	—
Pentachlorobenzene	20	—	—	0.05	30	—
Hexachlorobenzene	—	1000	—	0.05	30	—

ORNL = Oak Ridge National Laboratory
CCME = Canadian Council of Ministers of the Environment
RIVM = Dutch National Institute of Public Health and the Environment
ANZECC = Australia and New Zealand Environment and Conservation Council
RC = Remediation Criteria (agricultural land-use context)
EIV = Ecological Intervention Value
EIL = Ecological Impact Level

It is therefore clear that, without full knowledge of their scientific basis and how relevant they are to a specific site, GVs should be used with extreme caution in determining whether land is contaminated.

8 Comments and Discussion

The requirement to use ecological risk assessment under the new contaminated land regulations will encourage practical and scientifically-based decision-making regarding the need for and scope of remedial action. This is to be welcomed since it will ensure that remediation will occur on the basis of risk, rather than hazard.

A criticism of the statutory guidance is that the amount of ecological damage required to constitute significant harm is extremely large. Expressed as an 'irreversible or other substantial change in the functioning of the ecological system within that location', it implies that few impacts short of wholesale habitat destruction will suffice, given that ecosystems are highly resistant to change. Toxic effects on individuals are effectively dismissed, with the focus being placed purely on the system as a whole. Both on moral grounds and in terms of potentially allowing continual and uncontrolled chemical loading into the natural environment, this is perhaps too lenient a definition.

In relation to the assessment of whether harm *is* occurring, it should be noted that the interface between science and legal proof is fraught with pitfalls and that a 'balance of probabilities' will be difficult to establish. In relation to the assessment of the *possibility of significant harm* occurring, while the use of ecotoxicological GVs and biological tests will help to achieve a rigorous appraisal of a site, there is a gap between what the new regime requires from ecological risk assessment and what it can actually deliver. This is particularly true in regard to the use of GVs, in the context of the definition of ecological harm described above, and the use of biological tests.

The outcome of the Environment Agency's research on ecological risk assessment presents an approach which is similar to the many tiered systems that are already in existence. However, there is a lack of detail given on Tiers 2 and 3 and, as described above, the relationship of the GV-based approach to the requirements of the statutory guidance remains unclear. Subjects which were not covered, and which perhaps should have been, include the conversion of GVs between classes of organisms and the use of detailed wildlife risk assessment models (see, for example, Suter *et al.*[5] and US EPA[6]).

On a final note, it remains to be seen how the ecological aspects of the new regime will work in practice. Given the amount of ecological harm required, the uncertainty surrounding ecological risk assessment and the ability of ecosystems to recover, it is the author's belief that land will only be remediated on ecological grounds where there are obvious and substantial impacts that have already occurred. Since, in the author's experience, there are few examples of land-based

[5] G. W. Suter, L. Barnthouse, S. Bartelle, T. Mill, D. Mackay and S. Paterson, *Ecological Risk Assessment*, Lewis Publishers Inc., Chelsea, Michigan, USA, 1993.

[6] US EPA, *Wildlife Exposure Factors Handbook*, Office of Research and Development, Washington DC, USA, 1993.

contaminants having caused such impacts, it is likely that ecological risk assessment will play a minor role, if any, in managing contaminated land under the new regime.

Remediation Methods for Contaminated Sites

PETER WOOD

1 Introduction

Risk assessment procedures undertaken on any contaminated site will establish a range of contaminants (hazards), pathways and receptors, will identify possible linkages of concern and will determine a set of remediation objectives. Any remediation subsequently proposed for the site must then be capable of fulfilling these objectives. In general terms the primary aims of any remediation is to break the linkage between contaminant source and the receptor to achieve a reduction of actual or potential risk so that unacceptable risks are reduced to acceptable levels.

Remediation can thus be achieved by one or more of the following:[1,2]

- elimination of the hazard by removal or treatment/modification of the contaminant
- control of the hazard by isolation or separation of the contaminant
- interruption of the pathway of contaminant movement and exposure
- protection or removal of the receptor (essentially involving an interruption of the pathway)

The need for, and extent of, any remediation will depend on the magnitude of the risk, which itself is a measure of the probability of an unwanted effect coming about and the consequence of it. A high probability risk with a severe consequence will raise considerably more concern than a low probability risk with a minor consequence.

[1] P.A. Wood, Remediation Methods for Contaminated Sites, in *Contaminated Land and its Reclamation*, eds. R.E. Hester and R.M. Harrison, Issues in Environmental Science and Technology No. 7, Royal Society of Chemistry, Cambridge, UK, 1997, pp. 47–71.

[2] P.J. Young, S. Pollard and P. Crowcroft, Overview: Context, Calculating Risk and Using Consultants, in *Contaminated Land and its Reclamation*, eds. R.E. Hester and R.M. Harrison, Issues in Environmental Science and Technology No. 7, Royal Society of Chemistry, Cambridge, UK, 1997, pp. 1–24.

Issues in Environmental Science and Technology No. 16
Assessment and Reclamation of Contaminated Land
© The Royal Society of Chemistry, 2001

P. Wood

Whereas there are numerous remediation technologies available,[1,4–6] the selected remediation method for any particular site would be dependent on site specific requirements. Any contamination can impact:

- the ground surface
- any debris or constructional materials
- substrates in the unsaturated zone
- soil atmosphere in the unsaturated zone
- substrates in the saturated zone
- groundwater where DNAPL, LNAPL, or dissolved phase contaminants can occur

The remediation process selected will thus need to be able to deal with the:

- specific type of contaminant and its occurrence—different contaminants might require different remediation processes
- type of substrate—remediation processes are not applicable to all substrates

Whereas a chapter[1] in an earlier issue of this series summarised the full range of remediation methods available, this chapter introduces overall remediation approaches and then details a selection of innovative approaches that are starting to be applied within the UK. Emphasis is on remediation technologies that are essentially *in situ* techniques and which generally involve minimal disturbance of the ground surface.

2 Summary of Remediation Approaches

There is a very wide range of remediation methods available to tackle contamination although three broad approaches can be distinguished:[1,3]

- engineering approaches—these are primarily the traditional methods of excavation and disposal to landfill, or the use of appropriate containment systems
- process based techniques include physical, biological, chemical, stabilisation/solidification, and thermal processes
- hydraulic measures and natural attenuation

Brief definitions of the main methods available within each of these categories are provided in Table 1, while the subsequent sections provide a summary of each.

[3] CIRIA, *Remedial Treatment for Contaminated Land*, ISBN 0-86017-408-5, CIRIA, London, UK, 1998, 12 vols.

[4] R. Armishaw, P. Bardos, R. Dunn, J. Hill, M. Pearl, T. Rampling and P. Wood, *Review of Innovative Contaminated Soil Clean-up Processes*, Report LR 819 (MR), ISBN 0-85624-677-8, Warren Spring Laboratory, Stevenage, UK, 1992, 231 pp.

[5] S. P. Barber, R. P. Bardos, H. C. van Ommen, J. J. M. Staps, P. A. Wood and I. D. Martin, *Contaminated Land Treatment: Technology Catalogue*, Report AEA/CS/16419078/036, AEA Technology, Culham, UK, 1994, 115 pp.

[6] E. Riser-Roberts, *Remediation of Petroleum Contaminated Soils*, ISBN 0-87371-858-5, Lewis Publishers, New York, NY, USA, 1998, 542 pp.

Engineering approaches
Landfill involves the three stages of soil excavation, transport, and burial at the landfill site. Contaminants in the soil are not necessarily removed, stabilised or destroyed on site and are ultimately transferred to another site. Landfills are designed to ensure that contaminants are either isolated from the environment or subjected to attenuation processes so that they no longer cause harm to the environment

Containment measures are those which are designed to prevent or limit the migration of contaminants, that may be either left in place or confined to a specific storage area, to the wider environment. Approaches include hydraulic measures, capping, the use of break layers and low permeability barriers amongst others

Process based techniques
Physical processes used in soil treatment are used to remove contaminants from the soil matrix, concentrating them in process residues that require further treatment or safe disposal. Contaminants in the concentrated fractions may subsequently be destroyed, recovered by some other process (*e.g.* chemical or thermal), or may be disposed of to landfill

Biological processes of soil treatment depend on the natural physiological processes of micro-organisms, such as bacteria and fungi, to transform, destroy, fix or mobilise contaminants. Phytoremediation processes can use higher plants to degrade contaminants, to fix them in the ground, to accumulate them in a harvestable biomass or to release them to the atmosphere through transpiration

Chemical processes in soil treatment systems are used to destroy, fix or neutralise hazardous compounds. Many processes in other categories may use chemical processes for the treatment of effluents and gaseous emissions

Stabilisation/solidification processes involve solidifying contaminated materials, converting contaminants into less mobile chemical forms and/or binding them within an insoluble matrix presenting a minimal surface area to leaching agents. It is when the process results in chemical fixation of contaminating substances that the term stabilisation can be applied

Thermal processes use heat to remove or destroy contaminants by incineration, gasification, desorption, volatilisation, pyrolysis or some combination or these

Hydraulic measures and natural attenuation
Hydraulic measures refer to the control of the groundwater regime so that a contamination source or contaminated groundwater is separated, isolated, treated or contained

Natural attenuation refers to the effective reduction of contaminant toxicity, mobility or volume by natural processes

	Advantages	Disadvantages
Table 2 Advantages and disadvantages of disposal to landfill[3,4]	Wide range of contaminants and materials can be disposed of	Contaminants are only removed to different location
	Contaminants are completely removed from a site	Quality of containment may be difficult to ensure
	Can be relatively low cost	Long term integrity not well understood
	Permits rapid remediation of site	Long term monitoring required
	Associated excavation offers advantages of improving ground conditions or recycling selected materials	Leachate and gas collection systems may be required

Landfill

Contaminated sites are frequently remediated by excavation of the contaminated material and subsequent disposal of this to a controlled landfill. The approach represents a rapid method of dealing with a contaminated site but it has been criticised as it represents only a transfer of the contaminated material from one location to another rather than a final solution. Relatively low landfill disposal costs are the major incentive for this type of disposal although recent trends are encouraging a movement away from this approach. The major advantages and disadvantages of removal to landfill are outlined in Table 2.

Contaminated material disposed of to landfill must be prevented from causing any further environmental damage. The principal approaches that contribute towards prevention are:[1,4]

- containment
- attenuation

Containment measures at a landfill are designed to isolate the disposed material from the environment such that any liquid or gaseous interchange is minimised or controlled. This isolation may be achieved by a range of containment techniques including lining, capping, cover systems and on occasion vertical barrier systems. Some of these containment techniques may also be used for the on-site disposal of excavated contaminated material which obviates the need for transport to a landfill. Control of any leachate or gaseous products may also be achieved by leachate or gas collection systems. The effective design and installation of a containment system requires extensive geological and hydrological investigation, modelling and monitoring. Although low permeability is a necessary characteristic of containment materials complete impermeability is rarely attained in practice. However, any materials used for containment may also act as a substrate for attenuation mechanisms. A further degree of containment can be achieved if the contaminated material is subjected to stabilisation/solidification techniques prior to disposal.

Attenuation occurs as a result of various mechanisms operating in the landfill which serve to minimise the movement and/or reduce the toxicity of contaminants (see section on Monitored Natural Attenuation, p. 129). Attenuation mechanisms can be physical, chemical or biological. Physical methods include the adsorption and absorption of contaminants, filtration, dilution and dispersion. Chemical methods of attenuation include acid–base interactions, oxidation, reduction, precipitation and ion-exchange. Biological methods of attenuation include aerobic and anaerobic microbial degradation.

Containment

The concept of containment as a method for dealing with contaminated ground is based on the use of low-permeability barriers to isolate the contaminated material, or any associated leachate or gaseous products, from the environment. The barriers can be constructed from natural or synthetic materials, or a combination of both, and can be placed over, under or around a contaminated area or pollution source. The technique can be used to isolate existing hazards such as a contamination source, to prevent the spread of contaminants from a disposal site such as landfill, or to isolate specially designed mono-disposal sites for contaminated soil.

Remediation of many contaminated sites has been achieved by covering the surface with clean material incorporating a low-permeability layer. Whereas this may reduce infiltration and form a physical barrier to the contamination it may not necessarily control adequately the movement of contaminants. In order to provide adequate control it may be necessary to use such cover systems in conjunction with vertical and horizontal in-ground barriers or cut-offs to achieve partial or total isolation of the site. This isolation, in its extreme, can involve the complete enclosure of a site within an impermeable barrier to inhibit the ingress or egress of leachate and gasses in all directions. The following types of containment will be examined:

- cover systems
- in-ground barriers

Cover Systems. A cover system consists of a singe layer, or succession of layers, of selected non-contaminated material that covers the area of contamination. The purpose of a cover system is to provide specific physical and chemical properties such that one or more of the following is achieved:[1,3–5]

- prevention of exposure of at-risk targets to potentially harmful substances. These targets might include humans, vegetation, animals, buildings and the environment
- sustained growth of vegetation
- control of infiltration to, and subsequently from, the site
- any geotechnical requirements are achieved.

Table 3 details the requirements of a cover system and indicates the purpose of the various components that might be incorporated.

The choice of materials suitable for use as a cover system is wide and would

Table 3 Requirements and components of a cover system[4,5]

Requirements	Component
Minimise toxicity in upper layers of soil	All component layers—to provide sufficient thickness of clean materials
Limit surface water percolation and minimise leachate	Top soil—to support vegetation
	Sub-soil—to support vegetation
Control vertical gas movement	Low permeability barrier layer—to prevent passage of water, gas or VOCs
Prevent capillary movement of contaminants	Buffer layer—to protect barrier layer
Prevent soil erosion and dust generation	Drainage layer/system—control drainage
Support vegetation	Gas control layer/system–gas control
Inhibit root penetration into contaminated layers	Filter layer—control the movement of particles
Improve structural properties to facilitate road or building construction	Break layer—to prevent capillary movement of contaminants
Improve aesthetics	

depend on the physical properties required by any particular component layer. Possible materials include:

- natural clays, sub-soil and soils
- amended soils incorporating materials like pulverised fuel ash, lime and sludges
- waste materials such as fly ash, slags, dredgings, sewage sludges
- synthetic membranes and geotextiles
- concrete, asphalt *etc.*

In practice economic factors, in addition to physical properties, will influence the material selected and a compromise between ideal and locally available materials might be necessary. In this way a cover system can represent an effective treatment procedure at reasonable cost.

Cover systems have been used widely within the UK for the reclamation of many types of contaminated sites such as gasworks, tarworks, waste disposal sites, metal mining areas, chemical works and tanneries. Many of these have been successful but, because the contaminants still remain on site, a cover system is not necessarily an effective long-term solution in those cases where contaminants are mobile. Cover systems do not necessarily control groundwater movements, gaseous emissions and odours and, following completion, long-term monitoring may be necessary.

Quality assurance is of vital importance during the installation of a cover system whatever its design and whatever the type of materials used in construction. It is essential that the desired properties of individual component layers are maintained over the entire area and depth. Integrity must not be compromised by inappropriate or poor quality materials. Careless or inappropriate use of heavy construction equipment can damage previously completed layers, particularly when synthetics are used. The major advantages and disadvantages or a cover system are identified in Table 4.

In-ground Barriers. In-ground barriers can be used to isolate, usually by

	Advantages	Disadvantages
Table 4 Advantages and disadvantages of cover systems[3,4]	Wide range of contaminants and materials can be isolated Can be relatively low cost Permits rapid remediation of the site Suitable for use on large sites where large volumes of material might be involved	Contaminants are not removed from the site Possibility for failure over long term Leachate and gas collection systems may be required Groundwater movements not necessarily controlled Long term monitoring may be necessary Possible long term restrictions on use of site Need to consider contingency liability and insurance implications

physical means, a contaminated mass of ground from the surrounding environmental or other targets. Low-permeability material may be introduced around or under the contaminated site, or methods incorporating some sort of physical, biological or chemical control of contaminant migration can be used (see Section 3, Reactive Barriers, p. 136). In-ground barriers can be placed around, above and below a contaminant source to achieve complete isolation; a method that has been termed macroencapsulation.[4] In practice, however, it is often difficult to ensure the continuity of any barrier, particularly if underground obstructions are present.

Vertical barriers can be can be constructed by various methods:[3,4]

- excavation—excavation barriers require the removal of material to form a trench followed by the backfill of the barrier material. The types of material that can be used include clays, clay—bentonite mixtures, concrete diaphragm walls
- displacement—displacement barriers do not require any excavation prior to excavation. Instead the barrier material or form work is inserted by force directly into the ground. Particular types of displacement barriers include steel sheet piling and vibrated beam slurry walls
- injection—barrier material may be injected under pressure, often incorporating a mixing process to mix the injected material with soil components *in situ*. Injection methods include chemical grouting, jet grouting, jet mixing and auger mixing.

The effectiveness and applicability of barrier methods vary according to the types and nature of contaminants present, physical conditions of the site and the design life of the barrier. The long-term integrity of barrier materials may, in some cases where specific contaminants are present, not be known. Also, at present there is only limited experience with the installation of some barrier types. For example, horizontal in-ground barriers have only been installed in a few instances. Particular problems with their installation include ensuring the

Table 5 Advantages and disadvantages of in-ground barriers[3-5]	*Advantages*	*Disadvantages*
	Wide range of contaminants and materials can be isolated	Contaminants remain on site
	Can be economic where large volumes of contaminated material are involved	Installation difficulties if underground obstructions present
	Can be used to control solid, liquid and gaseous contaminants	Difficulties in ensuring barrier continuity
		Limited understanding of long term integrity
		Long term monitoring may be necessary
		Possible long term restrictions on use of site
		Need to consider contingency liability and insurance implications

Table 6 Advantages and Disadvantages of physical processes[3-5]	*Advantages*	*Disadvantages*
	Some methods are already or are becoming established	Secondary waste streams may require treatment or disposal
	Potential to reduce volume of material requiring disposal or expensive treatment	Soils with high clay or peat content may be difficult to treat
	Wide range of contaminants treatable	Use of some solvents will have health and safety implications
	Wide range of materials treatable	Quality assurance measures needed, especially for *in situ* methods
	Some *in situ* methods require only little site disruption	Approval by regulatory authority may be needed
	Mobile plant available for some methods	

continuity of the barrier under the entire contaminated area. The main advantages and disadvantages of using in-ground barriers are listed in Table 5.

Physical Processes

Physical processes separate contaminants from uncontaminated material by exploiting differences in their physical properties (*e.g.* density, particle size, volatility, by applying some external force (*e.g.* abrasion) or by altering some physical characteristic to enable separation to occur (*e.g.* flotation). Depending on the nature and distribution of the contamination within the soil, physical processes may result in the segregation of differentially contaminated fractions (for example a relatively uncontaminated material and a contaminant concentrate based on a size separation) or separation of the contaminants (for example oil or metal particles) from the soil particles. Table 6 summarises the main advantages and disadvantages of physical processes.

The range of physical processes includes a diverse variety of methods that include both *in situ* and *ex situ* approaches. This variation has been classified into two main groups:[5]

- washing and sorting treatments
- extraction treatments

Washing and Sorting Treatments.   Washing and sorting treatments are commonly referred to as soil separation and washing. The main aim of the processes is to concentrate the contaminants into a relatively small volume so that the costs associated with disposal and further treatment are related only to the reduced volume of process residues.

Washing and sorting falls into two main categories:

- separation from the soil of those particles containing the contaminants by minerals processing techniques, exploiting differences in the properties of individual soil particles. The volume of contaminated material requiring further treatment or disposal is thereby reduced
- removal of contamination from particle surfaces by scrubbing or attrition processes, or its transfer into an aqueous phase by leaching using liquid extractants or steam. The contaminant rich leach liquor can then be treated as waste water

Washing and sorting treatments have been used in several countries, particularly Germany and the Netherlands, for the treatment of a range of soils and contaminants. Their use in the United Kingdom is restricted although they have been used on occasion.

Extraction Treatments.   Extraction treatments involve processes that remove the contaminants from soils by involving a mobilising and/or releasing process to remove the contaminant from the soil matrix. Three main categories of extraction treatments are:

- soil vapour extraction—*in situ* process where a vacuum is applied through extraction wells to create a pressure gradient that induces gas-phase volatile contaminants to flow through the soil to the extraction wells where they then become removed from the soil (see Section 3, Soil Vapour Extraction, p. 135)
- electroremediation—an *in situ* process where an electrical current is passed through an array of electrodes that is embedded in the soil. When the current is applied, movement of contaminants in the pore water towards the electrodes is induced by electrolysis, electro-osmosis and electrophoresis. The electrodes have porous housings into which purging solutions are pumped to remove the contaminants and bring them to the surface, The purging solutions are then pumped to a water treatment plant for contaminant removal
- soil flushing and chemical extraction—processes that use chemical reagents, solutions or steam to mobilise and extract contaminants from soils. Mobilisation refers to the release of dissolved contaminant ions from sorbed or precipitated forms in soils and may form part of both *in situ* soil flushing or *ex situ* chemical extraction treatments.

Biological Processes

The objective of a biological remediation process is the degradation of contaminants to harmless intermediates and end products. The ultimate aim is the complete mineralisation of contaminants to carbon dioxide, water and simple inorganic compounds. A large number of organic contaminants can be degraded by micro-organisms and most biological treatments attempt to optimise conditions for degradation by the naturally occurring indigenous microbial population. Achieving these optimum conditions may require control of temperature, oxygen or methane concentrations, moisture content and nutrients.

Biological degradation processes can be either *in situ* or *ex situ* and either aerobic or anaerobic. Biological treatments have considerable scope for integration with other remediation processes and are applicable to both contaminated soil and groundwater. An advantage of the simpler biological treatments is their potential to be cost effective although long treatment times may be necessary. The presence of certain contaminants such as pesticides or heavy metals, however, may inhibit the effectiveness of a biological treatment. An additional problem is the possible creation of more hazardous intermediate products.

A range of biodegradation processes is available and methods of classification vary. Main groups include:

- *in situ* processes—involving the injection of air or water to convey oxygen and possibly nutrients into the underground contaminated mass
- dynamic *ex situ* processes—following extraction dynamic processes, in addition to the control of water, nutrients *etc.* are applied to mix the soil and encourage rapid degradation. Processes include landfarming, windrow turning and bioreactors
- static *ex situ* processes—following extraction the material is left undisturbed for the duration of the treatment, possibly under a liner. The condition of the material can be monitored and possible addition of water, nutrients and air made. The main static process is soil heap bioremediation, sometimes referred to as biopiles and composting
- phytoremediation—these processes use higher plants to degrade contaminants, to fix them in the ground, to accumulate within plant tissue or to release them to the atmosphere (see Section 3, Phytoremediation, p. 133).

Table 7 summarises the main advantages and disadvantages of biological processes.

Chemical Processes

Chemical treatment processes for the remediation of contaminated soil are designed either to destroy contaminants or to convert them into less environmentally hazardous form. Chemical reagents are added to the soil to bring about the appropriate reaction. In general, excess reagents may need to be added to ensure that the treatment is complete. This in turn may result in excessive quantities of unreacted reagents remaining in the soil following treatment. Heat and mixing may also be necessary to support the chemical reaction. Chemical processes can

Table 7 Advantages and disadvantages of biological processes[3-5]	*Advantages*	*Disadvantages*
	Applicable to both contaminated soil and groundwater	High cost of complex processes
	Potential for integration with other processes	May require long process times
	Simple processes can be cost effective	Possible formation of hazardous intermediate products
	High contaminant specificity is possible	Presence of some contaminants may inhibit degradation
		Most inorganic contaminants may not be treatable
		Some complex organic contaminants may not be treatable

Table 8 Advantages and disadvantages of chemical processes[3,4]	*Advantages*	*Disadvantages*
	Applicable to wide range of matrix types if good mixing/contact achieved	Effectiveness requires good mixing/contact which may be difficult with some soils
	High degree of chemical specificity possible	Unreacted chemical reagents may remain in the soil
		Any intermediate or by-products may be hazardous
		Pre-processing may be needed to remove debris, for size reduction or to form slurry
		Effective environmental control of *in situ* methods difficult

also concentrate contaminants in a manner similar to physical processes.

A range of chemical remediation processes is at various stages of development, both for *in situ* and *ex situ* applications. Many of these are based on the treatment of waste water or other hazardous waste. However, the range of processes that have been widely used at full scale is restricted. Major types include:

- oxidation–reduction
- dechlorination
- extraction
- hydrolysis
- pH adjustment

The major advantages and disadvantages of chemical remediation methods are summarised in Table 8.

An oxidation–reduction (redox) reaction is a chemical reaction in which electrons are transferred completely from one chemical species to another. The chemical that loses electrons is oxidised while the one that gains electrons is reduced. Redox reactions can be applied to soil remediation to achieve a reduction of toxicity or a reduction in solubility.

Oxidation and reduction processes can treat a range of contaminants including organic compounds and heavy metals. Oxidising agents that can be used include oxygen, ozone, ozone and ultraviolet light, hydrogen peroxide, chlorine gas and various chlorine compounds. Reduction agents that can be used include aluminium, sodium and zinc metals, alkaline polyethylene glycols and some specific iron compounds.

Chemical dechlorination processes use reduction reagents to remove chlorine atoms from hazardous chlorinated molecules to leave less hazardous compounds. Dechlorination can be used to treat soils and waste contaminated with volatile halogenated hydrocarbons, polychlorinated biphenyls and organochlorine pesticides.

Extraction techniques that can be used for the treatment of contaminated soil include organic solvent extraction, supercritical extraction and metal extraction using acids. The methods are applicable to soils, waste, sludges and liquids. Following extraction of the contaminant the extraction liquid containing the contaminant has to be collected for treatment.

Hydrolysis refers to the displacement of a functional group on an organic molecule with a hydroxide group derived from water. A restricted range of organic contaminants is potentially treatable by hydrolysis, although hydrolysis products may be as hazardous, or even more hazardous, than the original contaminant.

pH adjustment refers to the application of weak acidic or basic materials to the soil or groundwater to adjust the pH to acceptable levels. A common example is the addition of lime to neutralise acidic agricultural soils. Neutralisation can also be used to affect the mobility or availability of contaminants such as metals by enhancing their precipitation as hydroxides.

Stabilisation/Solidification

Stabilisation/solidification methods operate by solidifying contaminated material, converting contaminants into a less mobile chemical form and/or by binding them within an insoluble matrix offering low leaching characteristics. These processes can be used to treat soils, wastes, sludges and even liquids, and a variety of contaminant types. However, the treatment of organic contaminants is generally more difficult and more expensive. Many of the reagents used for stabilisation/solidification are proprietary products. An added benefit is the improved handling and geotechnical properties of the treated product that might result compared with the original contaminated material.

Stabilisation/solidification processes have been applied both *in situ* and *ex situ*, the latter being both on and off site. With an *ex situ* approach it may be necessary to landfill the stabilised product if an alternative use or disposal option is not possible. A disadvantage here is that the volume of the stabilised product can be considerably greater than the original contaminated material because of the quantities of stabilisation materials that have been added.

Stabilisation/solidification processes can be classified according to the type of material used as the binder. The more frequently applied methods use:

Advantages	Disadvantages
Applicable to inorganic and organic contaminants, although organics are less proven Applicable to soils, sludges and liquids Improved handling and geotechnical properties possible Rapid treatment possible *Ex situ* methods relatively easy to apply	Contaminants contained rather than destroyed or detoxified Increase in volume of material following treatment Some processes produce heat which can cause gaseous emissions Quality assurance measures needed, especially for *in situ* methods Uncertainties over long term performance, especially with organic contaminants Long term monitoring required

- Portland cement
- pozzolanic materials such as fly ash
- lime
- silicates
- clays (often used in conjunction with other materials)
- polymers (often used in conjunction with other materials)
- other proprietary additives

Table 9 summarises the advantages and disadvantages of stabilisation/solidification.

Thermal Processes

A wide range of thermal processes are at various stages of development[1,3-5] although the number of technologies that are commercially available is considerably more restricted. Techniques under development and commercially available can be either *in situ* or *ex situ*. Three *ex-situ* techniques will be outlined that operate in different temperature regimes:

- thermal desorption
- incineration
- vitrification

The advantages and disadvantages of thermal processes are summarised in Table 10.

Thermal desorption involves the excavation of the contaminated soil following by heating to temperature in the region of 600 °C. At these temperatures the volatile contaminants are evaporated and subsequently removed from the exhaust gasses by condensation, scrubbing, filtration or destruction at higher temperatures.[3] Following treatment it may be possible to re-use the soil depending on the temperatures used and the concentration of any residual contamination. Thermal desorption has its primary use in the treatment of organic contamination although it has also been used for the treatment of mercury-contaminated soils.

Table 10 Advantages and disadvantages of thermal processes[3-5]

Advantages	Disadvantages
Potential for complete destruction of contaminants	High cost of some methods due to high energy requirements
Applicable to a wide range of soil types although may be handling problems	Soil may be destroyed by high temperatures
Established technologies with some mobile plant for selected process types	Heavy metals contaminants may not be removed and become concentrated in ash
Possible re-use of soil if process temperature is not excessive	Potential for generation of harmful combustion products
	Control of atmospheric emissions required including condensing of volatile metals
	Approval by regulatory authority may be required

Incineration involves the heating (either directly or indirectly) of excavated soil to temperatures of 880–1200 °C to destroy or detoxify contaminants. Incineration can also be used for the treatment of contaminated liquids and sludges. Incineration results in the destruction of the soil texture and removes all natural humic components. The residues may also have high heavy metal contents. Exhaust gasses need to be treated to remove particulates and any harmful combustion products. A range of methods of incineration are available although the use of rotary kilns is probably the most widespread. Costs of treatment are heavily dependent on the water content of the material being treated and any calorific value that the material may have.

Vitrification involves the heating of excavated soil to temperatures in the region of 1000–1700 °C. At these temperatures vitrification of the soil occurs forming a monolithic solid glassy product. Contaminates will either be destroyed or trapped in the glassy product. The technology works by melting the alumino-silicate minerals in the soil which, on cooling, solidify to form the glass. In soils or waste where there are insufficient alumino-silicates these can be added in the form or glass or clay. The product from vitrification may have very low leaching characteristics. Exhaust gases require treatment for the removal of any volatile metals or hazardous combustion products. Vitrification is an expensive process and likely to be restricted in use for particularly hazardous contaminants that are not readily treated by other methods.

Hydraulic Measures and Natural Attenuation

Hydraulic measures for contaminant control involve the manipulation of the groundwater regime so that a contaminant source or contaminated plume is separated, isolated, treated or contained.[3] The main components of a system employing hydraulic measures include:

- extraction wells—which are used to lower the water table by withdrawing

groundwater
- injection wells—which are used to re-inject water into the ground (after treatment if necessary) or as part of a hydraulic system to mange plume migration or water balance
- infiltration trenches—used mainly for aquifer recharge

The pump and treat operations that are becoming more frequently used to treat or contain contaminated groundwater can be considered as a form of hydraulic measure.

Monitored Natural Attenuation involves the monitoring of the subsurface environment and processes to evaluate whether natural attenuation processes are sufficient to ensure an adequate reduction of contaminant concentration by the time the migrating contaminated groundwater reaches a receptor (see Section 3, Monitored Natural Attenuation, below).

3   Selected Innovative Approaches

This section aims to review *in situ* approaches that might be used to remediate contaminated ground with minimal disturbance of the ground surface. A wide range of suitable remediation technologies have been developed to tackle specific occurrences of contamination, types of contamination, and types of substrate. While many are proprietary variations of particular approaches there is a more restricted range of generic technologies. Table 11[7] provides a summary of the more established technologies that might be used for the *in situ* treatment of contaminated ground. However, it should be realised that the distinction between *ex situ* and *in situ* is not always easily recognised.

The remainder of this chapter introduces a selection of remediation processes that do not involve excessive disturbance of the ground surface or the bulk excavation and removal of contaminated material. Those considered are:

- monitored natural attenuation
- phytoremediation
- soil vapour extraction
- reactive barriers

Monitored Natural Attenuation

Attenuation processes have long been recognised as contributing to a reduction of contaminant concentrations in groundwater as it migrates away from a contamination source. Indeed, attenuation processes have been recognised as important controlling factors governing the acceptability of those landfill

---

[7] P. A. Wood and R. Swannell, Response to Spills Involving Ground Penetration and Contaminated Land Remediation, in *Interspill 2000: A New Millennium—A New Approach to Spill Response*, ISBN 0-85293-3177, Institute of Petroleum, London, UK, 2000, pp. 164–196.

Technology	To treat	Comments
Table 11 Selection of generic *in situ* treatment technologies[7] Containment	Substrate atmosphere and groundwater	Installation of low permeability barriers to reduce migration of soil vapours and contaminated groundwater
Soil vapour extraction	Substrate atmosphere	Removes VOCs from soil atmosphere by vacuum pumping and encourages further volatilisation from soil particles and groundwater surface
Bioventing	Substrate in unsaturated zone	Aerates soils by air injection to encourage biodegradation of contaminants. Can be augmented by addition of nutrients
Soil heating	Substrate in unsaturated zone	Heat is applied underground to thermally desorb contaminants from particle surfaces or increase volatilisation rates of volatiles/semi-volatiles. Heating mechanisms include hot air, steam and radio-frequency heating
Bioremediation (ground)	Substrates in unsaturated zone	Range of methods to encourage biodegradation of contaminants at any depth from surface to groundwater table. Can involve addition of nutrients, oxygen sources and water
Soil flushing	Substrate in saturated zone	Reagents are injected into the soil to remove contaminants from particles and into cleaning solution. Solution is then pumped to surface
In situ chemical treatment	Substrate in unsaturated zone and groundwater	Addition of reagents to achieve a chemical change in the contaminant to make it less toxic or less mobile/bioavailable. Approaches include oxidation/reduction, neutralisation, hydrolysis
Pump and treat	Groundwater	Removes contaminated groundwater by pumping for treatment on the surface. Treated groundwater can subsequently be returned

Table 11 *(cont.)*

Technology	To treat	Comments
Air sparging	Groundwater	Air is pumped through groundwater to air-strip volatile contaminants which are then removed by soil vapour extraction
Biosparging	Groundwater and substrate in saturated zone	Addition of air/oxygen and nutrients below groundwater to encourage biodegradation of contaminants. Also brings about air sparging and potentially improved biodegradation in the substrate in the unsaturated zone
Bioremediation (groundwater)	Groundwater	Addition of nutrients and/or oxygen sources into contaminated plume to encourage biodegradation
Reactive barriers or *in situ* reactors	Groundwater	Introduction of reagents into the ground to treat contaminated plume as it migrates through the barrier or treatment zone (includes bioremediation of groundwater)
LNAPL recovery	LNAPL free product	Pumping of free product to surface by a variety of methods
DNAPL recovery	DNAPL pools	Pumping of DNAPL to surface

operations that rely on attenuation and dispersion for contaminant control,[8] or for environmental protection in the event of landfill containment failure.[9] In more recent years natural attenuation (or intrinsic remediation) has been increasingly shown to be a potentially viable approach for the control and treatment of contaminated groundwater, and the protection of water resources,[10-12] especially those contaminated with fuel hydrocarbons, chlorinated solvents, and other organic compounds.

Natural attenuation is defined by the US Environmental Protection Agency as

[8] DoE, *Landfilling Waste*, ISBN 0-11-751891-3, Department of the Environment, Waste Management Paper No. 26, HMSO, London, UK, 1986, 205 pp.

[9] R. N. Young, A. M. O. Mohamed and B. P. Warkentin, *Principles of Contaminant Transport in Soils*, ISBN 0-444-882936, *Developments in Geotechnical Engineering 73*, Elsevier, New York, NY, USA, 1992, 327 pp.

[10] P. V. Brady, M. V. Brady and D. J. Borns, *Natural Attenuation: CERCLA, RBCA's and the Future of Environmental Remediation*, ISBN 1-56670-302-6, Lewis Publishers, New York, NY, USA, 1998, 181 pp.

[11] P. A. Wood, Natural Attenuation and RBCA, *Land Contam. Reclam.*, 1999, **7**, 231–235.

[12] B. C. Alleman and A. Leeson, *Natural Attenuation of Chlorinated Solvents, Petroleum Hydrocarbons, and Other Organic Compounds*, ISBN 1-57477-074-8, Batelle Press, Columbus, OH, USA, 1999, 378 pp.

Table 12 Summary of natural attenuation processes[1,8-10,13-16]	*Physical*	*Chemical*	*Biological*

Physical	*Chemical*	*Biological*
Dilution	Gas dissolution/exsolution	Aerobic degradation
Dispersion	Precipitation	Anaerobic degradation
Colloidal mechanisms	Redox reactions	
Adsorption		
Absorption		
Filtration		

'the biodegradation, dispersion, dilution, sorption, volatilisation, and/or chemical and biochemical stabilisation of contaminants to effectively reduce contaminant toxicity, mobility, or volume to levels that are protective of human health and the ecosystem'.[10]

Many processes combine to achieve natural attenuation and these can be grouped under the three main categories of physical, chemical, and biological. Table 12 indicates a selection of the main processes.

The application of monitored natural attention as a cost effective technology for contaminated site treatment involves the prediction and monitoring, on a site-specific basis, of *in situ* attenuation processes that might bring about a reduction in contaminant concentrations. As such, the application requires a detailed knowledge of the chemistry and physico-chemical conditions within the site, including soil and groundwater conditions, hydrogeology, and biological activity. The procedure also requires site assessment mechanisms and extensive monitoring systems for successful application.[17] Brady *et al.*[10] provide an excellent summary of a comprehensive protocol described by Wiedemeier *et al.*[18] for the evaluation of natural attenuation.

Natural attenuation should not be considered as a do nothing option. It requires considerable site assessment and monitoring procedures while potentially providing significant health and environmental benefit. Furthermore, the attenuation processes can result in real reductions in contaminant mass as well as a reduction in concentrations. A proper evaluation of natural attenuation on a site specific basis, coupled with appropriate risk evaluation procedures, can result in potential savings in remediation costs.

A case study of the monitored natural attenuation of explosives is provided by Zakikhani *et al.*[19] The contaminant source was a lagoon that had been used for

[13] Environment Agency, *Natural Attenuation of Petroleum Hydrocarbons and Chlorinated Solvents in Groundwater—Phase 2*, 1998, Http://www.environment-agency.gov.uk/modules/MOD53.360.html

[14] L. J. Thibodeaux, *Environmental Chemodynamics: Movements of Chemical in Air, Water and Soil*, ISBN 0-471-61295-2, John Wiley & Sons, New York, NY, 1996, USA, 593 pp.

[15] USEPA, *A Citizen's Guide to Natural Attenuation*, Report EPA 542-F-96-015, USEPA, Office of Solid Waste and Emergency Response, USA, 1996, 4 pp.

[16] K. H. Tan, *Principles of Soil Chemistry*, ISBN 0-8247-8989-X, Marcel Dekker, New York, NY, USA, 1993, 347 pp.

[17] NNAGS, Network on Natural Attenuation in Groundwater and Soil, 1998, Http://www.shef.ac.uk/uni/projects/nnags/aims.htm

[18] T. H. Wiedemeier, in *Intrinsic Bioremediation*, eds. R. E. Hinchee *et al.*, Batelle Press, Columbus, OH, USA, 1995, pp. 31–52.

[19] M. Zakikhani, J. C. Pennington, D. W. Harreison and D. Gunnison, Monitored Natural Attenuation of Explosives at the Louisiana Army Ammunition Plant, *Land Contam. Reclam.*, 2000, **8**, 233–239.

the disposal of explosives-laden waste water. As explosives can be naturally attenuated by microbial mineralisation, microbial transformation, and/or immobilisation, the study set out to establish an appropriate monitoring regime and so determine whether attenuation was occurring. Contamination levels were found to be generally static to declining over time with a declining contamination mass. It was considered that the site's capacity for natural attenuation was low but that nevertheless monitored natural attenuation is a viable option that should be considered amongst the options available for other explosives contaminated sites.

Phytoremediation

Phytoremediation is a technology still under development. Although some investigations to date suggest that the technology shows considerable potential for both metals and organic contaminants there have not been many extensive applications of the approach.

Phytoremediation methods can be applied to a achieve range of remediation functions for both metal contaminants and organic contaminants:[20,21]

- phytoextraction—phytoextraction refers to the uptake of metal contaminants in the soil by plant roots and the translocation of these into the above ground portions of the plant. After a period of growth the plants are harvested and the contaminants thus removed from the site. Phytoextraction thus acts to remove the contaminant source.
- rhizofiltration—rhizofiltration refers to the adsorption or precipitation of metal contaminants in solution onto plant roots, or their absorption into the plant roots. As this approach is concerned primarily with contaminants in solution it can be used to address groundwater problems. Rhizofiltration thus acts to remove contaminants from the groundwater and so prevents their migration by groundwater flow processes.
- phytostabilisation—phytostabilisation is the use of plants to immobilise metal contaminants in the soil and the groundwater through absorption and accumulation by roots, adsorption on to roots, or precipitation within the root zone of plants. The process reduces the mobility of the contaminants and prevents migration to the groundwater and also reduces bioavailability for entry into the food chain. Phytostabilisation thus acts to interrupt the contaminant pathway from the soil to the groundwater.
- phytodegradation—phytodegradation is the breakdown of organic contaminants external to the plant through the effect of compounds produced by the plant (such as enzymes) or the breakdown of organic contaminants taken up by plants through their metabolic processes. Complex organic contaminants are broken down into similar molecules which are incorporated within the plant tissue.

[20] USEPA, *A Citizen's Guide to Phytoremediation*, Report EPA 542-F-98-011, USEPA, Office of Solid Waste and Emergency Response, USA, 1998, 6 pp.

[21] B. D. Ensley, Rationale for the Use of Phytoremediation, in *Phytoremediation of Toxic Metals: Using Plants to Clean Up the Environment*, eds. I. Raskin and B. D. Ensley, ISBN 0-471-19254-6, John Wiley & Son, New York, NY, USA, 2000, pp. 3–11.

Table 13 Plants with potential for phytoremediation[20,22–24]

Plant	Contaminant	Medium
Sunflowers	Radionuclides	Groundwater
Hybrid poplars	Chlorinated solvents/nitrates	Groundwater
Hybrid poplars	Heavy metals/atrazine	Soil
Grasses	Petroleum/heavy metals	Soil
Indian mustard	Radionuclides	Soil
Turnips	Heavy metals	Soil
Willow	Radionuclides	Soil
Cottonwood	Chlorinated solvents	Groundwater
Mulberry	PAH	Soil

- rhizodegradation—rhizodegradation is the breakdown of organic contaminants in the soil through microbial activity that is enhanced by the presence of the plant root zone. Certain micro-organisms can digest organic contaminants as a source of nutrition and energy and break them down into harmless products.
- phytovolatilisation—phytovolatilisation refers to the uptake of an organic contaminants by plants and their subsequent release by transpiration to the atmosphere in the original or a modified form.
- hydraulic control—hydraulic control refers to the use of plants whose roots reach down to the water table and which take up large quantities of water. Such plants are capable or removing sufficient groundwater to reduce flow rates from a contaminated site. Hydraulic control measures thus act to interrupt the contaminated groundwater flow pathway.
- restoration—plant species that are tolerant to the contamination, whether or not they have been selected to achieve one or other of the phytoremediation approaches indicated above, can be used to restore sites which otherwise might be devoid of vegetation. The plants thus have the potential to decrease the potential migration of contaminated soil particles by wind erosion and surface runoff, and to reduce leaching of soil contamination to the groundwater. Restoration thus acts to interrupt the contaminant pathway to environmental receptors and the groundwater. Such a vegetation cover would also help to minimise some of the exposure pathways for humans.

A range of plants species have the potential to achieve phytoremediation processes and can be used for a variety of contaminants. Table 13 summarises a selection of these.

Phytoremediation has been used to treat sites contaminated with a variety of contaminants including heavy metals, solvents, PAHs, PCBs, hydrocarbons, radionuclides, explosives, and pesticides.[20,21,24] The approaches are probably

[22] H. Black, Absorbing Possibilities: Phytoremediation, *Innovations*, 1995, **103**, 6.

[23] R. Levine, *Phytoremediation: Natural Attenuation that Really Works*, 1997, http://www.em.doe.gov/tie/spr9712.html

[24] M. C. Negri and R. R. Hinchman, The Use of Plants for the Treatment of Radionuclides, in *Phytoremediation of Toxic Metals: Using Plants to Clean Up the Environment*, eds. I. Raskin and B. D. Ensley, ISBN 0-471-19254-6, John Wiley & Son, New York, NY, USA, 2000, pp. 107–132.

most suited to sites with only low to medium contamination where they might be used for polishing (final treatment). As the depth to which remediation can be achieved is generally restricted to the depth of the root zone of the plant, the technology is best suited for areas of shallow contamination. The use of larger plants such as trees would enable sites with deeper contamination to be treated, although concern has been raised about adverse effects of drawdown on the water table with some tree species (assuming that some form of hydraulic control is not the objective).

Consideration should also be given to the consequences of any contaminant accumulation in the plant components. Two possible concerns are transfer to the foodchain as a result of ingestion of plant parts by insects, rodents, or larger herbivores, and the transfer of contamination to the surface layers or wider environment *via* leaf fall or wood gathering. Ensley[21] outlines the rationale for the use of phytoremediation while Dickinson *et al.*[25] discuss issues associated with planting trees on contaminated soils and provide some guidelines. Dickinson *et al.* conclude that tree planting may be particularly beneficial to the breakdown of organic contaminants, for the manipulation of heavy metal loadings, and in developing healthy sustainable soils at brownfield sites.

Soil Vapour Extraction

Soil vapour extraction (SVE) is a technique which extracts contaminants from the soil in vapour form.[26] Consequently, SVE systems are used to remove gases and organic volatile or semi-volatile contaminants from the unsaturated zone. Figure 1 provides a schematic of a SVE system. The process works by applying a vacuum through a series of underground extraction wells which draws up soil gases for surface treatment. In addition to extraction wells a series of air injection wells might be installed to actively or passively encourage air flow through the soil and so improve contaminant removal rates. An additional benefit of introducing air into the soil is that the SVE process can be better controlled and that bioremediation of some contaminates might be stimulated.

In some SVE schemes the injection wells might be extended below the water table so that injected air passes through contaminated groundwater. In this way volatile contaminants in the groundwater can be air stripped and bioremediation again encouraged. This process is termed air sparging.

SVE is a considerably more established technology compared with monitored natural attenuation, phytoremediation, or reactive barriers and has been used on many occasions. The largest treatment area in the UK was for the treatment of the Millennium Dome site at Greenwich, UK.[27] This site was one of a former gasworks and benzene plant and the range of contaminants included BTEX in

[25] N. M. Dickinson, J. M. Mackay, A. Goodman and P. Putwain, Planting Trees on Contaminated Soils: Issues and Guidelines, *Land Contam. Reclam.*, 2000, **8**, 87–101.

[26] USEPA, *A Citizen's Guide to Soil Vapor Extraction and Air Sparging*, Report EPA 542-F-96-008, USEPA, Office of Solid Waste and Emergency Response, USA, 1996, 5 pp.

[27] D. L. Barry, The Millennium Dome (Greenwich Millennium Experience Site) Contamination Remediation, *Land Contam. Reclam.*, 1999, **7**, 177–190.

Figure 1 Schematic of a soil vapour extraction system

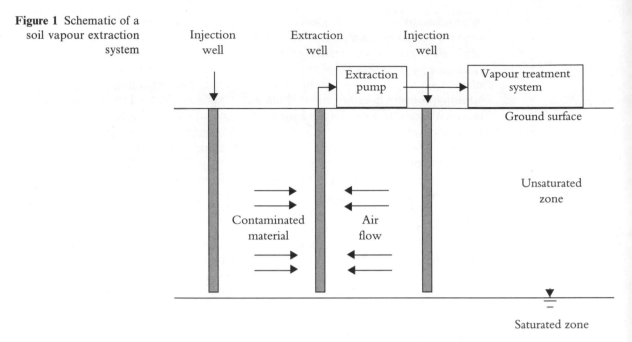

the soil and perched groundwater that extended over an area of 4.3 ha. A dual phase SVE system was installed to treat this area which consisted of 300 wells. The system involved a series of groundwater wells to draw down the water table and remove dissolved and free phase contaminants, and a series of vacuum extraction wells to remove volatile organic compounds. It was necessary for the extraction wells to be closely spaced as the presence of any subsurface structures would reduce efficiencies and also because the time available for remediation was restricted. The extensive vacuum caused by the system created a corresponding ingress of air through the shallow groundmass and thus also encouraged biodegradation of the organic contaminants. The system successfully removed 110 tonnes of volatile organic compounds or which 12 tonnes were benzene. These quantities are believed to represent about 70% of the volatile organic compounds and nearly 95% of the benzene that was available for remediation. It is estimated that treatment costs by SVE were some six times lower than an alternative excavation approach. Additional environmental benefits were identified as a reduced loss of benzene to the atmosphere and no traffic impact on already congested roads in the area.

Reactive Barriers

Reactive barriers are essentially engineered structures installed underground to treat dissolved phase contamination in the groundwater that passes through it. The structures can be permanent, semi-permanent, or replaceable. These relatively novel structures have also been called treatment walls, treatment

Table 14 Reactive barrier processes used for specific contaminants[29,31-33]

Contaminant	Reactive barrier/process
Lead/acidity	Precipitation on and neutralisation with limestone
Halogenated hydrocarbons	Redox reactions using iron
Chromate	Reduction using iron
Phosphate	Adsorption and precipitation using iron
Aromatic compounds	Aerobic reactions and biodegradation encouraged by oxygen releasing compounds
Chlorinated solvents	Enhanced biodegradation
Heavy metals	Sorption on zeolites
White spirit (petroleum hydrocarbons)	Biodegradation by air sparging and sorption on peat

curtains, and permeable barriers.[28-31]

The reactive barriers can be installed across the path of a migrating contaminant plume to prevent further plume growth, or constructed immediately downstream of a contaminant source to prevent plume development. As the contaminant plume passes through the barrier the contaminants are either removed by the barrier and immobilised or are transformed into less harmful compounds. A range of treatment processes can be used within reactive barriers to treat both organic and inorganic contaminants.[29,31] These include:

- sorption
- precipitation
- oxidation/reduction/ion exchange
- biodegradation

Table 14 indicates some of the approaches that can be used for specific contaminants.

Reactive barriers can be installed through the entire depth of an aquifer and keyed into a low permeability layer, or installed in only the upper portion. Figure 2 indicates a schematic of a permeable reactive barrier. These are normally

[28] R. C. Star and J. Cherry, In situ Remediation of Contaminated Groundwater: the Funnel and Gate System. *Ground Water*, 1994, **32**, 465–476.

[29] USEPA, *A Citizen's Guide to Treatment Walls*, Report EPA 542-F-96-016, USEPA, Office of Solid Waste and Emergency Response, USA, 1996, 5 pp.

[30] J. A. Cherry, S. Feenstra and D. M. Mackay, Concepts for the Remediation of Sites Contaminated with Dense Non-Aqueous Phase Liquids (DNAPLs), in *Dense Chlorinated Solvents and other DNAPLs in Groundwater*, eds. J. F. Pankow and J. A. Cherry, Waterloo Press, Guelph, Canada, 1996, pp. 475–506.

[31] F.-G. Simon and T. Meggyes, Removal of Organic and Inorganic Pollutants from Groundwater Using Permeable Reactive Barriers: Part 1—Treatment Processes for Pollutants, *Land Contam. Reclam.*, 2000, **8**, 103–116.

[32] M. J. Baker, D. W. Blowes and C. J. Ptacek, Phosphorous Adsorption and Precipitation in a Permeable Reactive Wall: Applications for Wastewater Disposal Systems, *Land Contam. Reclam.*, 1997, **5**, 189–193.

[33] T. F. Geurin and T. McGovern, A Funnel and Gate System for Remediation of Dissolved Phase Petroleum Hydrocarbons in Groundwater, *Land Contam. Reclam.*, 2001, **9**, in press.

Figure 2 Schematic of a permeable reactive barrier

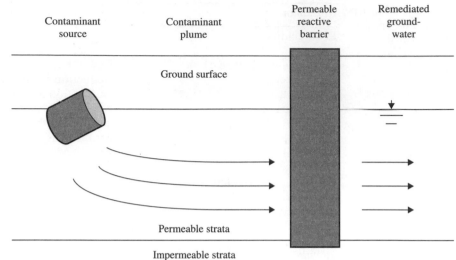

constructed by introducing reactive agents into the ground by either backfilling of an excavated trench or through injection *via* boreholes. The reactive agents are normally mixed with a suitable substrate to encourage high permeability, and are specifically selected for their ability to treat the contaminants concerned.

Permeable reactive barriers can also be used in conjunction with low permeability cut-off walls in a funnel and gate system.[28] The cut-off walls act as a funnel system to modify flow patterns and direct contaminated groundwater to the permeable reactive barrier that in turn acts as a gate. Meggyes and Simon[34] provide a summary of construction methods for reactive barriers and associated cut-off walls. Successful applications require a detailed understanding of groundwater behaviour and careful modelling to understand the impact of installing the low permeability barrier.[33,35]

Jefferis *et al.*[36] provide a case study of a permeable reactive barrier installed at an industrial site in Belfast, UK. Historic spillages had resulted in contamination of the groundwater at the site with chlorinated solvents. Concentrations of trichloroethene, trichloroethane and tetrachloroethene reached $390\,\mathrm{mg\,l^{-1}}$, $600\,\mu\mathrm{g\,l^{-1}}$, and $100\,\mu\mathrm{g\,l^{-1}}$ respectively. Reductive chlorination of these contaminants using iron filings in a permeable barrier was considered as a potential treatment option. The reactive barrier and associated cut-off walls had to be specifically designed to overcome site constraints. The size of the site was insufficient for

34 T. Meggyes and F.-G. Simon, Removal of Organic and Inorganic Pollutants from Groundwater Using Permeable Reactive Barriers: Part 2—Engineering of Permeable Reactive Barriers, *Land Contam. Reclam.*, 2000, **8**, 175–187.

35 G. Teutsch, J. Tolksdorff and H. Schad, The Design of In Situ Reactive Wall Systems—A Combined Hydraulic-Geotechnical-Economic Simulation Study, *Land Contam. Reclam.*, 1997, **5**, 125–130.

36 S. A. Jefferis, G. H. Norris and A. G. Thomas, Developments in Permeable and Low Permeability Barriers, *Land Contam. Reclam.*, 1997, **5**, 225–231.

treatment to be achieved by the normal approach of horizontal flow through the permeable barrier. Consequently, the system involved collecting groundwater and diverting it to the reactive barrier where vertical flow was induced, thus permitting a longer retention time within the barrier for a given ground surface area. Sampling showed that all chlorinated solvents were rapidly degraded upon entering the reactive barrier, and that after a travel distance of 6 m concentrations of trichloroethene dropped to $< 100 \mu g^{-1}$.

4 Summary and Conclusions

This chapter has introduced the overall remediation approaches available for the treatment of contaminated sites and provides a summary of selected generic *in situ* techniques. It has detailed a selection of innovative approaches that are starting to be applied within the UK. Emphasis has been on remediation technologies that are essentially *in situ* techniques and which generally involve minimal disturbance of the ground surface.

Whereas many remediation approaches show promise, their careful selection and application are paramount to a successful outcome. Monitored natural attenuation is being increasingly considered as a potentially viable approach for the control and treatment of contaminated groundwater, and for the protection of water resources. Phytoremediation is a technology still under development. Although the technology shows considerable potential for both metals and organic contaminants there have not been many extensive applications of the approach. Soil vapour extraction is a considerably more established technology and has been used on many occasions for the treatment of volatile organic compounds. Reactive barriers are relatively novel structures that can be used to treat a range of contaminants in a groundwater plume as it migrates through the barrier.

Legal Liabilities and Insurance Aspects of Contaminated Land

ANTHONY J. LENNON

1 Introduction

We have finally got the new liability regime (Part IIA of the Environmental Protection Act—inserted into that Act by Section 57 of the Environment Act 1995) for contaminated land throughout the majority of the United Kingdom (UK). What we do not seem to know is how much contaminated land there is throughout the UK and what the significance is of the contamination that is or may be present. Various studies have been made to determine the amount of contaminated land there is in the UK, and if the results of the various studies are to be believed the liabilities associated with this land are massive. A study[1] in 1998, for example, found that 58 000 hectares of brownfield land was unused or potentially available for development. In another study,[2] it was found that at least 30% of petrol stations contaminated groundwater.

What is contaminated land, what are some of the liabilities and how can the liabilities be managed so that they have a minimal effect on a company's financial strength are some of the issues that will be considered throughout this chapter.

2 Contaminated Land

What exactly is contaminated land? There is quite obviously a definition contained within the modified Environmental Protection Act 1990 (EPA90) (section 78A(2)), which can be summarised as any land that is causing significant harm to health of living organisms, interference with ecological systems or where there is a significant possibility of such harm or interference being caused. In addition any land that is or is likely to cause pollution of *controlled waters* is also to be considered as contaminated land.

This statutory definition is very strict and probably only relates to a small sub-set of all land that may have some contamination in it. Such lesser contamination may not be an immediate threat to the environment, as catered for

[1] *The ENDS Report*, 2000, No. 311, p. 1.
[2] A review of current MTBE usage and occurrence in groundwater in England and Wales.

Issues in Environmental Science and Technology No. 16
Assessment and Reclamation of Contaminated Land
© The Royal Society of Chemistry, 2001

by the strict definition, but may nonetheless restrict the uses to which the land can be put to, or may impede future development of the land.

We therefore, have at least two types of contaminated land. Land which is so heavily contaminated that it falls within the provisions of the statutory definition could be shown as of immediate interest and designated as 'CONTAMINATED LAND' and land which has impaired development potential and could be designated as 'contaminated land'.

The higher order 'CONTAMINATED LAND' incurs statutory penalties and is subject to the provisions of the EPA90—notices, inspections, remediation requirements, *etc.*, whereas the lower order 'contaminated land' is usually only an issue when land is to be developed or when land is subject to transfer from one party to another.

For the purposes of this chapter the distinctions between 'CONTAMINATED LAND' and 'contaminated land' will be used throughout.

3   Statutory Provisions in the UK

Many detailed descriptions have been made about the new statutory regime for 'CONTAMINATED LAND' in the UK, and the reader can refer to these or preceding chapters of this publication to obtain a detailed description of the provisions. All that I would like to stress is that the system is essentially one that is likely to be used as a last resort in circumstances where land is a real environmental problem and remedial works need to be undertaken in a timely manner. The provisions should not be taken out of context and should not be considered as if they apply to all areas of contamination, since they are reserved for only the most serious of polluted sites.

The primary purposes of the statutory provisions are to:

(i)   Define the regulators with responsibility for 'CONTAMINATED LAND'
(ii) Provide powers to regulators to seek out 'CONTAMINATED LAND'
(iii) Define the liabilities of those that find themselves associated with 'CONTAMINATED LAND'
(iv) Provide powers to the regulators to manage 'CONTAMINATED LAND'

It should be noted that there are no provisions for dealing with anything other than 'CONTAMINATED LAND', *i.e.* there are no provisions relating to the handling of contamination during the normal course of land development, since it is assumed that this will be dealt with by either contractual issues at the point of land sale or during land development with the financial gains obtained in the land development funding the costs associated with the contaminated land. There is so little regulation of contaminated land that the only time there is any obvious attempt at regulating the quality of land is during the planning process when land is developed.

The conclusion that can be drawn from the above is that there are many forms of land that may have some form of pollution contained within or under it, but only a small proportion of such land is subject to statutory control and hence is unlikely to cause significant problems to land owners except when:

- They wish to develop the land,
- They wish to sell the land or business,
- They wish to raise a mortgage on the land,
- They wish to use the land as security for a pension fund, *etc.*

4   Sale of Land/Businesses

Although not exclusively, it is often when businesses are sold that there are concerns raised about potential environmental liabilities. It seems to matter little whether or not there is real tangible evidence that the liabilities will materialise. What seems to be the key is the interest the parties have in proceeding with the sale and the potential use of environmental liabilities as a negotiating position within the sales process. However, in a number of cases there are real tangible fears associated with environmental liabilities and often the parties associated with transfer of businesses look to insurance to solve their problems for them.

The concerns may be associated with (a) the potential for the new landowner to face a regulator's action for remedial works to be undertaken on the site to remove suspected contamination, or (b) fears that known or suspected contamination may have migrated away from the site and be threatening to cause damage to third parties or their interests.

An attempt is often made within a sale contract to aportion liability between the parties involved in the sale, irrespective of what statutory liabilities there may be. Although not an exhaustive list of options available, sale contracts for land or businesses can have some of the following features:

(a) Seller retains full liability for all claims arising from pollution conditions pre-existing as of the point of sale,
(b) Purchaser assumes liability for all claims associated with pre-existing pollution conditions and agrees to hold the seller harmless,
(c) Often there is a distinction made between known (or suspected) pollution conditions and unknown conditions that lead to a liability after sale completion,
(d) Purchaser holds seller harmless for any liabilities associated with continued operations of the site(s),
(e) Seller has decreasing liability for the environmental liabilities based on the number of years past the anniversary date of the sale agreement,
(f) Occasionally there are no explicit conditions relating to environmental liabilities and in such circumstances the contractual liabilities for environmental issues have to be considered in the context of all other liabilities dealt with by the sale agreement,
(g) There is frequently no distinction made between on-site remediation expenses associated with regulator action and those associated with development of the site [*i.e.* no distinction between 'CONTAMINATED LAND' and 'contaminated land' (see above)].

The ways in which potential environmental liabilities are handled within a sale contract are very variable and apart from reflecting the views of the companies involved also reflect the astuteness of the various legal advisors. Care, therefore, needs to be taken in land transactions to fully understand the imposed liability regime of the contract as well as just being concerned about the statutory regime for 'CONTAMINATED LAND' which may only apply in a very few situations.

The third party liabilities, in particular, are very difficult to quantify and will not be based on any obvious statutory regime. Risk will be dependent on the following:

(a) What is or has been undertaken on the site,
(b) The sensitivity of the site with repect to the local conditions,
(c) How the various processes were operated, the way chemicals were handled, *etc.*
(d) The local geology and hydrogeology of the area and how natural local conditions can influence the transport of contaminants within the environment.

Frequently, costs associated with the removal of known or suspected contamination from a site can be assessed and be quantified to a reasonable degree of accuracy; however, third party damage is not so easy to quantify or predict and can have immense effects on the residual liabilities associated with land ownership or occupation.

5   Principles of Insurance

Insurance is concerned with financial risk transfer. It is an arrangement whereby one party (the insurer) promises to compensate another party (the insured or policyholder) a sum of money if an event occurs which causes the insured to suffer a financial loss. The responsibility for paying for such financial losses is then transferred from the policyholder to the insurer. In return for accepting the burden of paying for losses when, and if they occur, the insurer charges the insured a price—the insurance premium.

The presumption, based on statistical analysis by the insurance company, is that if there are enough people purchasing the class of insurance the total pool of premium generated is greater than the cost of the total liabilities encountered during the insured period. Insurance companies can also balance losses on one class of insurance with profits from another, thereby balancing books in any one accounting period. Insurers can also gain income from investing the premium payments and hence earning income on the premiums before any claims may be paid.

Major problems come when the total losses incurred by an insurance company exceed the available income together with any reserves they may have built up over a period of years. This can arise because of a number of reasons:

- more of the insured risks occurred than first estimated,
- the insurance company did not understand fully the risk they were accepting,
- the financial consequences of the risk were far higher than had been estimated by the insurers,

- the liability regime understood by the insurer changed from underwriting the risk to when the damage occurred,
- the insurer was held liable for costs they had not anticipated, *etc.*

Some of the above reasons have caused the insurance industry to be very circumspect about environmental liabilities and their ability to assess effectively these liabilities in a profitable manner. If the insurers cannot underwrite risks profitably, then they will not have enough funds to pay claims and hence the insured, when the risk occurs, may find that the insurance company is unable to meet its financial liabilities.

Insurance works only because insurers can collect together large numbers of similar exposures or risks. Since the markets work with such large numbers of exposures, they can employ statistics to the claims experience and hence work out the probabilities of claims occurring. The larger the number of similar risks that are insured the more accurate the insurers can be in calculating the likely loss or claim rates and hence developing premiums that provide them with profit and yet are equitable and more truly reflect the probability of the risk occurring and a claim being made against their policies.

The general insurance providers, however, do not apparently have the experience or expertise to consider environmental risks in the same way and hence seek to exclude environmental claims from a variety of insurance policies. The question is why and what is being done to improve the situation for potential insureds?

Claims Made vs. Occurrence Wording Insurance Policies

There are two main alternative operating principles for insurance policies—*claims made or occurrence wording*. An understanding of these two different operating principles is vital to be able to appreciate the insurance industry's concern over environmental liabilities and their apparent inability to assess properly their exposures.

Occurrence wording: This type of policy format works on the principle that the insured purchases a policy for a fixed period of time (say one year) and the policy will then respond to any claims arising out of actions of the insured that occurred during the period of insurance—no matter how many years later the damage manifests itself.

Claims made: This type of policy format relies on the loss that the policy is providing indemnity for to occur during the insured period and for the claim to be lodged against the insurance policy during the insurance period.

Public liability insurance provides indemnity to the insured against their legal liability incurred for property damage and bodily injury to third parties. It normally provides indemnity to the insured for the financial consequences of damage to third parties caused by any activities of the insured giving rise to such

145

damage, which would not be insured under more specific forms of insurance. Such policies in the United Kingdom are normally written on an occurrence wording basis. Since pollution damage can take many years to manifest itself, any indemnity policy issued on an occurrence wording basis is almost impossible to assess for potential pollution liabilities because of this potential massive time delay between potential exposure and the manifestation of damage.

6   History of Environmental Insurance

As with so many things relating to environmental protection we have to look to the United States of America for the first real experience the insurance industry had in considering environmental risks. Until the late 1970s the American insurance markets offered public liability insurance on an occurrence wording format with no pollution restrictions. Frightened by the problems associated with asbestos and the ever-tightening legislative system in the USA, the market attempted to restrict cover for pollution related damages.

Initial attempts were made to place restrictions on claims that were related to gradual pollution. It was soon discovered that for all practical applications it is extremely difficult to differentiate between gradual pollution damage and more instantaneous or sudden and accidental pollution damage. As claims against the partially restricted policies started to flow, it became apparent that the USA courts were not making consistent decisions and interpretation of the pollution exclusions. The consequence of inconsistency and the protracted litigation cases between the insurers and the insureds caused the market to respond by placing total pollution exclusions on their public liability policies (known in the USA as Comprehensive General Liability policies (CGL)).

The imposition of pollution restricted CGL policies allowed the development of pollution specific insurance products that could be underwritten independently of other forms of insurance.

USA Experience

The development of environmental policies was not without its problems. By the end of the 1970s there were only two insurers able to offer such policies. By the end of 1983 there were some 40 insurers offering environmental policies. Unfortunately by the end of 1984 the environmental insurance market collapsed leaving only one insurer surviving in the USA market.

The USA market failed so spectacularly because the underwriters did not hold true to their principles and apply sound technical principles when underwriting the risks. The lure of potentially large premiums and the pressure to make environmental insurance easier to obtain caused short cuts to be made in underwriting the risks. After the collapse of the class of business in the USA, there was a steady growth of providers and policy types over the subsequent years. This has led to the market for environmental insurance in the USA exceeding an estimated $1 000 000 000 gross written premium per annum.

UK Insurance Response to Environmental Liabilities

UK insurers' initial response to the problems of environmental liabilities and their effects on them (the insurer) were largely to ignore them and pretend that they did not exist. This lack of realisation has now started to change and in 1991 the general UK insurance market clarified the cover they were providing, by restricting cover to sudden and accidental incidents leading to pollution related damage by employing exclusion wording similar to:

'This policy excludes all liability in respect of pollution or contamination other than caused by a sudden identifiable unintended and unexpected incident which takes place in its entirety at a specific time and place during the Period of Insurance'.

Pollution Exclusion

The 'gradual pollution exclusion' has now become almost universal on all public liability policies, but what does it mean? Unfortunately the UK insurance industry has learnt nothing from the experience they had in the USA during the 1980s and has fallen into the trap of thinking that it can restrict effectively its exposure to environmental liabilities by limiting cover to sudden and accidental incidents. It has been learnt in the USA that the concept of sudden and accidental pollution damage is very difficult to interpret effectively and consistently. It is not clear from the words used what cover is being provided. A simple example will be sufficient to illustrate the point:

An explosion attributable to methane gas from a landfill site is quite obviously a sudden and accidental incident. Unfortunately, however, in deciding a claim the insurance markets will not look for the obvious cause of any damage, but will seek to find the root or proximate cause of the damage. In this example the proximate cause could be attributed to any of the following.

- gradual build up of methane in the space prior to ignition,
- gradual decomposition of the waste in the landfill,
- gradual seepage of methane from the landfill,
- gradual deposit of waste in the landfill.

As can be seen most of the above could be considered as gradual (or often expected or capable of anticipation) and hence would probably be excluded from cover from a UK public liability insurance policy. The issue in this case is not actually which of the causes would be considered as the proximate cause, it is rather that there are a number of potential causes, most of which may be considered as gradual and that the exclusion wording is ambiguous and allows interpretation. The other point to consider in the above exclusion wording is that it relates to an incident leading to damage and not the damage itself.

Although a few years ago there were indications that the UK insurance market was going to impose a total pollution exclusion on public liability insurance policies, no concrete signs have developed over recent times. Some general insurers have even started to provide insurance for environmental risks within standard public liability insurance policies. However, such cover is seldom as comprehensive as specific environmental insurance and can often exclude any indemnity for historical contamination.

7 Land Owners and Environmental Liabilities

Owners of land face potential liabilities with respect to environmental damage both from the past and present land uses. It is often thought, incorrectly, that the primary liability facing landowners is that of associated past land use. Past land use does give rise to substantial potential liabilities, although after the House of Lords[3] decision concerning the case of Cambridge Water *vs.* Eastern Counties Leather, the civil liability land owners may face for past contamination may not be so great as was initially thought.

This case involved the pollution caused to groundwater and the liabilities of the polluter (Eastern Counties Leather—ECL) after it was found that the groundwater could not be used for an authorised use (potable water source by Cambridge Water—CW). The final decision of the House of Lords, on appeal, was that the defendant (ECL) could not be held liable for the damages arising to CW from the solvent-contaminated groundwater. This was based on the view that ECL could not have foreseen that the solvent spilled over many years of operation would have migrated downwards into the groundwater (rather than evaporating) and caused the groundwater impairment that led to CW being unable to use the water for drinking water purposes.

Although potential liabilities for damages caused by contaminated land and past industrial activities have been addressed by the House of Lords in this one case, the costs of cleaning contaminated land and the acceptability of that land once cleansed to potential land purchasers is still a very real issue.

What should be of equal concern to land owners, particularly those that lease land to others (for example industrial estates), is the potential for the site to be contaminated by current land uses. Linked with these are the liabilities that may fall on landowners for contamination to land and damages to third parties as a result of today's activities rather than historical uses.

8 Pollution Liabilities for Contaminated Land Sites

It must be appreciated that contaminated land sites can exhibit pollution liabilities of two basic forms. There can be liabilities arising from pollution to people and property off site—so called third party liabilities—and also liabilities relating to pollution of the site itself and the consequential need to clean up the site. The second liability may, and in some cases is likely to, lead to damage to people or property off site.

As discussed above, damage caused to a third party's interests from contamination migrating from a contaminated site is unlikely to be covered by a current public liability policy. Any migration would almost certainly be judged by an insurer to

---

[3] Cambridge Water Company *v.* Eastern Counties Leather plc (1993).

be gradual in nature. In this situation the only hope for the owner of the land is either:

- have a specific pollution liability policy for the site, or
- hope that the migration can be traced back to a time when the current owners had a public liability policy without pollution exclusion and hope that the insurance policy was written on an occurrence worded basis. As time progresses, however, this is likely to become increasingly difficult because of the difficulty in being able to link the pollution damage with an incident that occurred away in the mists of history.

Costs to clean up the insured's own sites are not normally available from the traditional insurance market. It is not a claim that would be considered under a current public liability policy and if the contamination was known about, and also very severe then the site is unlikely to obtain site-specific environmental insurance to cover the required clean-up costs. The only situation where there may be a glimmer of hope from an insurance perspective is in situations where those facing liability for clean-up were tenants on the land and not the landowners themselves. It was, and still is, relatively common practice for tenants to have their public liability policies extended to provide indemnity to them for any damage they caused to the freeholder's land or property. If such a policy extension were to be available on an old occurrence worded policy held by a tenant the property damage part of the cover may provide them with some opportunity to gain an amount of insurance indemnity for any requirement placed upon them to clean up their landlord's land.

Most landowners and tenants hold property policies, which provide indemnity to the insured for the costs of rebuilding the premises as a result of fire or other named perils. These policies also provide indemnity to the insured for debris removal on site arising out of a specified named peril, *e.g.* a fire. These policies, although providing a degree of own site clean-up cover, would only respond to contamination and debris removal arising out of the named peril and would, therefore, not respond to the general removal of contamination on the insured's premises. In such circumstances, insurers may also wish to distinguish between debris removal (bricks and rubble) and contamination of the ground.

9 Current Manufacturing Activities

Manufacturing industry today is over-regulated and hence there is no potential for manufacturing industry to give rise to either land contamination or damages to third parties. Is this a valid comment? In my view it is most certainly not for a number of very important reasons:

- Not all potentially polluting processes are regulated.
- Regulation is still largely based on media interests, *e.g.* water pollution, air pollution, *etc*. Because of this it is quite conceivable that over-regulation for one environmental medium may cause a transfer of pollution potential to another environmental medium. To some extent this has been addressed under the principle of Integrated Pollution Control, but this form of

regulation only encompasses a very small proportion of manufacturing activities in the UK.

- Regulation tends to concentrate on normal operations in a manufacturing activity and does not always address potential aberrations in operational practices.
- The sanction in the event of a failure to comply with legislation is usually no more than a fine and very rarely does it encompass compensation or other means of providing restitution to injured parties or the environment.

For all the above reasons and more, I believe that regulation in the UK and throughout the rest of the world does not guarantee that a regulated manufacturing process will never cause damage to the environment. In the 1998 Annual Report of the Environment Agency,[4] it was reported that there were over 17 800 reported water pollution incidents in England and Wales during the reporting period. Although this number is likely to include a large proportion of relatively minor incidents, every one of them could have the potential of causing significant damage to the aquatic system and hence damage to other people's interests and property.

One simple example of where current manufacturing activities have given rise to contamination will suffice to demonstrate that today's activities can and do cause pollution and contamination. A small property in North London was owned by a company that let the site out as an industrial estate. One of the units was let to a metal electroplating company. The company went bankrupt and stripped all the valuable materials and equipment from the site one night and left the land owner with the problem of disposing of all waste materials that were left, together with a significant pollution legacy. The walls of the building were so badly corroded with plating chemicals that it would have been possible to push a hand through the brick wall without encountering any significant resistance from the brickwork. The floor and underlying ground were also grossly contaminated by the plating processes that had only ceased operation a matter of days before. The landowner was faced with removing and cleansing his site since he was unable to obtain any restitution from the occupiers of his industrial unit. There was also the problem of whether or not the contamination had affected the integrity of any surrounding units that he owned.

It must be appreciated, recognised and always remembered that manufacturing activities that are assumed to be benign are anything but that, and that they have as much potential for causing damage to the environment as the predecessor processes operating in Victorian or earlier times.

10   Environmental Insurance

As has been mentioned before, Public Liability policies in the main have some form of pollution exclusions attached to them. Property Damage policies, which provide indemnity in the event of damage being caused by named perils, only provide indemnity for removing contamination from that portion of a site

---

[4] *Water Pollution Incidents in England and Wales 1998*, the Environment Agency, ISBN 0-11-310150-3.

actually affected by the named peril (if at all). For example, a large site which has a fire in one building, yet causes contamination over the whole site, will not have indemnity to pay for the clean-up associated with the contamination from the fire from any land other than that immediately surrounding the fire. It is very unlikely that a large site contaminated by a fire would have any adequate insurance indemnity available to decontaminate it from pollution caused by a fire or other perils. Because existing policies have significant restrictions placed upon them in terms of the degree of cover provided for pollution related liabilities and clean-up costs, there have been developed two particular policies to indemnify insureds for these liabilities.

Pollution Legal Liability Insurance

This provides indemnity to the insured for damage caused to third parties as a result of pollution conditions emanating from the insured's premises. It has been used extensively in the USA and is now available in the UK. Some companies refer to this type of policy as Environmental Impairment Liability insurance. This title is a slight misnomer because it does not indemnify the insured for environmental damage *per se*, but for bodily injury and property damage caused as a result of pollution caused by the insured.

Own Site Clean-up Insurance

Within the UK, as in the USA, the pollution regulators have a series of powers available to them now to require landowners and occupiers to remove contamination from their sites, from land and from controlled waters.

An insurance policy is available that will indemnify the insured for the costs incurred, up to the limit of indemnity, in the event of any pollution regulator using their powers to institute a clean-up of their site. This is a particularly important policy since it is quite common for a third party pollution claim to lead to a clean-up requirement on the site giving rise to the problem and the current public liability policies will not provide any indemnity for such expenses.

Composite Environmental Insurance Policies

Over recent years there has been a move to combine coverages under one policy form and to allow the purchasers of the insurance the opportunity to select what type of insurance they wish to purchase from a list of options that may be available. In its very raw form, insurers have effectively combined the types of cover described above and placed them into one policy. Often there are extensions to this basic cover for such things as:

- business interruption costs for the insured in the event of them incurring such losses as a result of pollution damage they cause,
- legal defence expenses associated with defending such claims either from third parties or from a requirement of a regulator to undertake remediation on the insured's premises,

- indemnity for damage to the insured's property arising from pollution damage,
- indemnity to a waste producer (could be a land developer) for potential liability associated with the transportation of waste that, during its transport, causes pollution related damage, *etc.*

Some more specific environmental insurance policies that are now available for land development and land sales are now described.

Stop-loss Insurance

When a site is investigated by a consultant and a Phase II survey is undertaken it is usually possible to identify whether or not the land is contaminated and if so what type of remedial works will be required, either to placate regulators or to assist in the development of the site. Unfortunately, even with the best investigatory techniques that are available, no consultant can either guarantee that that all contamination has been found, or indeed that all types of contamination that are present have been looked for. This inability has led to potential problems when it comes to developing sites. In one case in the North West of England, a remediation project that was anticipated to cost £750 000 to complete ended up by costing in excess of £1 500 000. This led to severe financial hardship for the developer and potentially left the site in a more dangerous condition than it would have been if the works had never been started. Luckily, in this case the developer was able to find the necessary funds to undertake the final remediation, but if they had not then the site would have been left unsafe in a partial state of completion. This is one simple example of how projects can overshoot the anticipated costs and hence lead to uncompleted remedial works and financial ruin for developers.

Stop-loss insurance provides protection to those involved in land development such that they can insure for the potential extra costs associated with planned remediation. In order to obtain this insurance, considerable review of the site investigation reports and the cost estimation has to be undertaken in order to get an acceptable degree of comfort regarding the risks involved.

Normally the insured (the developer or contractor) is responsible for anticipated costs of remediation together with an additional sum (the buffer amount) in excess of the anticipated costs. After these amounts have been exhausted, the stop-loss insurance activates and indemnifies for any additional costs of remediation thereafter.

A variation of the above approach is for the developer/contractor to share in the available insurance. For example, it is not uncommon for the insured to trade a lower figure for the buffer amount against, say, a 5% share of the remediation expense in excess of the insured amount plus buffer.

A further variation of stop-loss insurance is on sites where there are no planned remedial works to be undertaken because, after due investigation, no evidence is found of any unacceptable contamination. Provided it can be clearly stated what would constitute unacceptable levels of contamination on such a site, insurance can be provided for the eventuality of unacceptable contamination being found during the course of a development, even though none was expected. Such insurance can be considered as providing indemnity for the 'development risk'. Again, as with standard stop-loss cover, the insurance would trigger for incurred costs of development in excess of an agreed amount.

Environmental Contractual Liability Insurance (Representation and Warranty Insurance)

Quite often in a contract relating to the sale of land there are general conditions that require the following:

- the seller to divulge all known information
- the seller to warrant that they have declared all information that is known to them
- the seller to warrant that the information declared is correct and does not contain any material inaccuracies
- the seller to assume liability for any failure in the warranties given
- the seller to assume liability for certain known or unknown environmental conditions.

Notwithstanding the liabilities that the seller may be forced to retain (see above) they will inevitably attempt to limit the financial value of any liabilities and if possible seek to limit the time in which they retain the liabilities. This is often in conflict with the seller's ideal situation of selling the land and not retaining any liability for past pollution conditions.

In such circumstances, the seller (or the purchaser if they are the party being required to give indemnities to the seller) can purchase contractual liability insurance to indemnify them for the residual retained liabilities arising from the sale agreement. Such insurance is relatively new and goes away from the traditional environmental insurance, which tends to relate to providing indemnity for certain prescribed liabilities. Contractual liability insurance on the other hand is designed to sit alongside the sale agreement and cover all liabilities arising from the agreement, except for those that have been openly excluded. The use of this insurance can provide additional comfort to the parties in a transaction because they do not have to check and see whether the wording in the sale agreement matches the declared areas of indemnity contained in the insurance policy. All that is needed is a review of the exclusions contained in the policy.

Such insurance is relatively new, but has proved to be very popular, both for the acquisition of land, and in particular for the sale and purchase of businesses and the associated land assets.

Benefits of Environmental Insurance

There are numerous reasons why such policies should be considered, not just for reclaimed contaminated land sites, but also for sites where contamination is not suspected. Some obvious reasons are:

- unknown past and future contamination/pollution damage covered,

- in investigations consultants may miss contamination,
- reclamation strategy may not be robust enough to deal with all found contamination,
- may be insufficient information on past land use,
- perceptions against contaminated land may harden,
- contamination may be present even though none found at investigation,
- ongoing land uses may cause additional contamination post reclamation,
- reclaimed contaminated sites that have benefit of site-specific environmental insurance may be more attractive to investors and end users,
- banks and other lenders can be added to the policies to note their interest in the property,
- land owners can also be added to the environmental policies to ensure that should their tenants at any time become bankrupt or if they ran away from their liabilities, there is at least some insurance indemnity available to provide the land owner with protection against the financial consequences of these environmental liabilities,
- can be written and provided on a multi-year basis.

11   Environmental Consultants

In the past, landowners and developers have been happy to rely upon the advice given by their environmental consultants in order to provide them with comfort that the land they are interested in is safe and free from any undue liabilities.

The cast of a survey of a contaminated land site, together with associated sampling and analysis, may run into hundreds of thousands of pounds. The remediation of the site as a result of the consultant's sampling, analysis and advice may cost many millions of pounds. The value of the land, once it has been developed, may be many tens or indeed hundreds of millions of pounds. If the consultancy was negligent in the way in which it took the samples or caused the analysis to be undertaken, in their assessment of the results of the analysis, or in the way in which they proposed the site should be cleaned up, and as a result the site is faced with a further clean-up bill of millions of pounds, therefore making the site in the short term worthless, who is to pay?

It is an accepted fact that consultants owe a duty of care to their clients. Consultants' clients expect a quality of work above that which would normally be supplied by an average person with average knowledge. Consultants are experts in their chosen areas of work and therefore the client can expect a quality of work commensurate with that professional standing and position. Unlike the USA, however, this duty of care is only shown to the consultant's primary client and the UK consultant does not normally need to consider any further users of their work. In order to limit their liability American consultancies tend to place limitations of liability either within standard contract conditions or within their reports, making it quite clear that their reports should be utilised only by their primary client, and for the original purpose of the consultant's brief.

In order for a client to successfully seek damages from a consultant, they must first prove that the damages they have faced were as a result of the consultant's negligence. This must be proven on the balance of probabilities. It is quite conceivable, however, that contamination may well have been missed not because the consultant was negligent but because of unsuspected contamination that could not possibly have been suspected or found by the consultant.

12   Collateral Warranties

In order for consultancies to allow their reports to be utilised by people other than their primary clients, it is not uncommon in the UK for consultants to enter into what is known as a 'collateral warranty'. For example, a developer may wish to purchase a plot of land and utilise a consultant to identify any potential environmental issues. The consultant produces a report and on the basis of the report the developer purchases the land. Having developed the site the developer then wishes to sell the land to a third party, but the third party does not wish to have a survey of the site undertaken for his own benefit, but wishes to rely upon the developer's initial survey. In such circumstances the developer, the third party purchaser and the consultant enter into a collateral warranty, thereby allowing the third party purchaser rights to the consultant's report. It is often the case that the consultant's client will wish the warranty to impose additional liabilities on the consultant beyond those associated with normal levels of professional service.

Insurers have to be happy that the collateral warranty is worded in such a way as to provide them with an adequate degree of cover or protection in the event of the third party taking any claims for negligence against the consultant. If entered into blindly collateral warranties can extend significantly the exposure of insurers in the UK to claims that would normally be outside the scope of common law or contractual responsibilities. But does a collateral warranty have any real value? A collateral warranty is only as good as the insurance that is in effect at the time of any claim being made, rather than being dependent upon the consultant's insurance that is in force when the collateral warranty is drawn up. With the potential for professional indemnity insurance in the future being limited such that it will not cover any claims relating to pollution or contamination, any collateral warranty entered into in the past may have its future value seriously undermined.

Limitation of Liability

A consultant can be subject to two forms of liability: both to liability under contract law and for claims relating to negligence and brought under tort.

In the USA it is very common for consultants to attempt to limit their liability through conditions in their contracts. The liability is often limited to the value of the contract undertaken or some other amount specified within the contract conditions. There has been a degree of debate about whether or not a condition limiting liability in a contract would also limit liability if a claim were taken for negligence under tort. It is now becoming more common for American environmental consultancies to place within their contract a condition limiting not only liability under contract, but extending that liability to any other claims for damages howsoever brought against the consultancy.

The practice of limiting liability in contracts is not common in the UK, although it is becoming more and more prevalent and is something that is not uncommon in consultants' contracts where the consultancy has any experience of work within the USA. The contract condition limiting liability has been tested at various times within the USA[5] and has been upheld particularly in situations where the contract value is only a very small proportion of the claim being brought against the consultant. This, for example, could be in a situation where a £25 000 consultant's report is used as the basis for a multi-million pound decision to purchase any individual plot of land.

Some insurers have used the type of wording found associated with public liability policies to restrict indemnity for sudden and accidental pollution only, to qualify indemnity for professional indemnity (PI) policies. Does the concept of sudden and accidental incident have any meaning when applied to a consultant's PI policy?

PI insurance with any form of pollution exclusion for environmental consultants is not worth the paper it is printed on. Any consultancy that is involved in environmental consultancy work must have the security of knowing that they do not have such an exclusion on their policy. It must be questioned whether a PI policy with any sort of pollution exclusion will provide a sufficient degree of indemnity should the worse happen and a claim is made against the insured and then hence against their PI insurance.

Environmental consultants tend not to have a very large capital base. Their greatest resource is their people. It is not uncommon for consultancies to rent out office space. In the event of a serious claim against them they can easily face insolvency. The provision of properly worded professional indemnity insurance offering cover for all negligence and associated claims is the only way a consultancy can operate effectively in today's market place.

13 Contractors

Consultants are engaging themselves in contracting services. This is particularly apparent where a consultancy undertakes a project management role on sites that require decontamination. Should there be any damage caused to the site or surrounding neighbours as a result of the consultant's contracting services, then they would be relying upon their public liability policy to provide them with the necessary level of indemnification. As is quite common, of course, their public liability policy will have a gradual, if not a total, pollution exclusion. If the damage is related to any pollution conditions arising out of any of the contracting services provided by the consultant, then it is possible that they will not have the necessary degree of insurance protection to cover the cost of the claims.

It is not uncommon for contractors to be held liable for any damage as a result of their activities. Where contracts are dealing with hazardous substances their clients almost certainly require them to assume responsibility for damage caused whilst handling or dealing with the hazardous materials. How can contractors accept such liability with a public liability policy that contains pollution exclusion?

There is now the opportunity for contractors to obtain specific pollution

[5] For example in 1992, California Court of Appeals in MARKBOROUGH CALIFORNIA INC. v. SUPERIOR COURT, 227 Cal. App 3d 705, 277 Cal Rptr. 919 (4th Dist. 1991).

indemnity insurance to supplement their public liability policies. In essence these policies provide indemnity to contractors should their activities, as a result of pollution conditions, give rise to property damage or bodily injury to third parties. The main features of these policies are that they are:

- sudden and gradual coverage for third party liability including defence costs,
- claims made policy; retroactive date provision,
- three-year extended reporting period option,
- completed operations coverage available,
- the insured's clients may be covered as additional insureds under policy,
- covered operations must be specified on the policy.

Where consultants are involved in providing contracting services as well as consulting services, they must ensure that they have an adequate degree of insurance to cover both aspects of their work. Although not common, it is possible to obtain tailor-made policies that encompass the indemnification provisions of both a professional indemnity policy and a public liability policy providing the necessary degree of cover for pollution related claims.

14 An Example

Although the location and details of the following example have been changed in order to provide confidentiality, the example is based on a real case.

Company X has two manufacturing sites within its overall group of manufacturing activities which it decides are outside its core interests and wishes to sell to raise capital to fund future development. It is planned that the capital produced in the sale will allow the acquisition of other manufacturing activities that are more akin to the core activities. Part of the proposed acquisition costs of the new manufacturing units will be funded by loans.

Company X has been operating on the two sites for at least thirty years and, although it is no longer used, has used a particularly contentious manufacturing process on one of the sites. Both sites are involved in processes that generate strong liquid effluents that were recycled, but which were stored in large unlined lagoons. During the past thirty years there have also been the inevitable spillages and leakages of liquid chemicals and also the storage of coal. Coal was used for a coal fired boiler which is no longer is use, since the conversion of all boilers to oil firing in the early 1980s.

Although Company X was aware that the main identified contentious process had left a legacy of contamination in the building which housed the process, they were caught unawares when the potential purchasers and their environmental advisors identified the following real problems:

- Costs associated with decontaminating the building which housed the contentious process,
- Costs associated with removing the associated contamination from the localised ground and groundwater under the site,
- Potential liabilities associated with the migration of the associated

contamination off site and its adverse effect on surrounding property, environment and people (the third party risk),

- On-site clean-up and third party liabilities associated with the spreading of boiler ash on part of the site which was subsequently used to house a process building which was still in use. The known contamination was primarily heavy metals, particularly arsenic,
- On-site clean-up and third party liabilities associated with known areas of additional contamination, and finally
- On-site clean-up and third party liabilities associated with any unknown areas of contamination.

The purchaser became very concerned over the potential liabilities associated with the above and sought to negotiate a sale contract which left all of the liabilities with the seller. The seller, although not wanting to retain any liabilities associated with their old sites after the sale, was sufficiently enthusiastic for the sale to go ahead that they agreed to retain some of the liabilities.

The sale of the two sites went ahead on the basis that the purchaser would share on a 50/50 basis with the seller all liabilities for the decontamination and on-site remedial works associated with the contentious process, but that the seller would retain all associated third party liabilities. The purchaser also agreed to accept liability for any on-site remedial works associated with the known or reported contamination. The seller, however, retained full liability for all third party damage and also for any remedial works that involved any contamination that was not declared during the sale process.

The sale went ahead and both parties were happy. It was agreed that no on-site remedial works would be undertaken in connection with the contentious process unless both parties agreed or until the regulators required such action to be undertaken. It was felt that this process building was stable and although it was clearly a potential source of damage there were no pathways for the contamination to migrate and cause damage as defined in the legislation. The seller also felt that the remaining liabilities were such that they could live with them on the basis that they would have had these liabilities if they had still been in possession of the two sites.

All went well until after the sale and the original seller had their accounts audited in preparation for their annual report. The auditors recognised the value of the sale and reviewed the sale contract to identify any provisions that may have to be made within the report and accounts. To the seller's horror, their auditors after reviewing the sale agreement identified that there were residual potential environmental liabilities associated with the sites, in some cases lasting for in excess of ten years. The auditors insisted that the seller show a provision on their report and accounts of at least £60 000 000 to cover the residual potential liabilities arising from the sale agreement. The seller could see that if such a large provision was shown in their accounts for these contingent environmental liabilities it could:

- Depress their share price,
- Have an adverse effect on their ability to borrow money in order to support

their proposed acquisitions,
- Leave them in a vulnerable position for take-overs, *etc.*

Having been forced to make the provision on their accounts the seller then sought to identify ways in which they could offset their potential liabilities and hence return their accounts to a more favourable position.

The seller, in considering the various options that they had before them to deal with the provisions contained within their accounts investigated the availability of insurance. The sale contract was very specific on which liabilities were to be retained by the seller and which were to be passed on to the purchaser. Unfortunately, although the remedial costs for on-site contamination were fairly well understood and quantified, the liabilities associated for third party damage were not, and it was these liabilities that the auditors were concerned about.

An insurance package was developed for the seller, such that all relevant liabilities associated with the sale were covered by insurance. Insurance was used to remove the uncertainty of the potential liabilities and hence effectively remove their effects from the accounts.

This is a simple example of how insurance can be used to provide a financial solution to a client's problems. The insurance was not purchased primarily as protection device for the purchaser; rather it was used as a mechanism to aid potential liability management. This is significantly different from the normal uses of environmental insurance as detailed above.

15   Conclusion

Public liability policies are already having their scope reduced so there is no intended cover for gradual pollution damage and very real possibilities that their scope will be further reduced in the near future to remove cover for all forms of pollution related damage. UK industry should be aware of this restriction and companies should consider whether or not their existing portfolio of insurance provides adequate protection.

Developers of contaminated land should understand the relationship they have with environmental consultants and should not consider the contract they have with the consultants as a surrogate for clean-up insurance.

There are now specific insurance policies available for industry and owners of land to indemnify them for a wide range of their potential environmental liabilities. These policies can be readily applied to development sites once they have been cleaned up to underwriter's satisfaction. For decontaminated contaminated sites, location-specific environmental insurance can remove some of the uncertainties from the site and promote further investment and development.

Subject Index